Pitman Research Notes in Mathematics Series

Submission of proposals for consideration
Suggestions for publication, in the form of outlines and representative samples, are invited by the Editorial Board for assessment. Intending authors should approach one of the main editors or another member of the Editorial Board, citing the relevant AMS subject classifications. Alternatively, outlines may be sent directly to the publisher's offices. Refereeing is by members of the board and other mathematical authorities in the topic concerned, throughout the world.

Preparation of accepted manuscripts
On acceptance of a proposal, the publisher will supply full instructions for the preparation of manuscripts in a form suitable for direct photo-lithographic reproduction. Specially printed grid sheets can be provided and a contribution is offered by the publisher towards the cost of typing. Word processor output, subject to the publisher's approval, is also acceptable.

Illustrations should be prepared by the authors, ready for direct reproduction without further improvement. The use of hand-drawn symbols should be avoided wherever possible, in order to maintain maximum clarity of the text.

The publisher will be pleased to give any guidance necessary during the preparation of a typescript, and will be happy to answer any queries.

Important note
In order to avoid later retyping, intending authors are strongly urged not to begin final preparation of a typescript before receiving the publisher's guidelines. In this way it is hoped to preserve the uniform appearance of the series.

Longman Scientific & Technical
Longman House
Burnt Mill
Harlow, Essex, CM20 2JE
UK
(Telephone (0279) 426721)

Titles in this series. A full list is available from the publisher on request.

John M Chadam

McMaster University, Canada

and

Henning Rasmussen

University of Western Ontario, Canada

(Editors)

Free boundary problems involving solids

Proceedings of the International Colloquium 'Free Boundary Problems: Theory and Applications'

Longman
Scientific &
Technical

Copublished in the United States with
John Wiley & Sons, Inc., New York

Longman Scientific & Technical
Longman Group UK Limited
Longman House, Burnt Mill, Harlow
Essex CM20 2JE, England
and Associated companies throughout the world.

Copublished in the United States with
John Wiley & Sons Inc., 605 Third Avenue, New York, NY 10158

First published 1993

AMS Subject Classification: 35, 47, 49

ISSN 0269-3674

ISBN 0 582 08767 8

British Library Cataloguing in Publication Data

A catalogue record for this book is
available from the British Library

Library of Congress Cataloging-in-Publication Data

A catalog record for this book is available

Printed and bound in Great Britain
by Biddles Ltd, Guildford and King's Lynn

Contents

Solidification I

Solidification II

Solid/solid phase transitions

Free boundary problems in fluid flow with applications
Fluid mechanics I
Fluid mechanics II
Porous flow I
Porous flow II
Geology and geophysics

Preface

This is the second of three volumes containing the proceedings of the International Colloquium "Free Boundary Problems: Theory and Applications", held in Montreal (Canada) from June 13 to June 22, 1990.

The Scientific Committee was composed of J. Chadam, F. Clarke, A. Friedman, P. Fife, H. Glicksman, H. Rasmussen and I. Stakgold.

The Organizing Committee consisted of J. Chadam, F. Clarke, M. Delfour, M. Goldstein, B. Ladanyi and H. Rasmussen. Invaluable assistance and direction was received from the International Committee composed of A. Fasano, M. Frémond, K.-H. Hoffmann, M. Niezgódka, J. Ockendon, M. Primicerio and J. Sprekels.

The Montreal meeting was the fifth in a series of International Colloquia (Durham (UK) 1978, Montecatini (Italy) 1981, Maubuison (France) 1984, Irsee (Germany) 1987). The fruitful exchange which characterized these previous meetings continued at the Montreal meeting. Pure and applied mathematicians, applied scientists and engineers gathered to share their results on free boundary problems arising in their disparate disciplines. A wide spectrum of physical, numerical and mathematical methods were added to the common pool of knowledge in the subject. There were approximately 175 individuals from 25 countries whose energetic participation was the key to the success of the meeting. We are grateful to them all.

We are pleased to have been offered the privilege of holding the meeting for the first time in North America on the 100th anniversary of Stefan's work on the formation of ice in polar seas. In his paper, published a year later in 1891 he compared the solution (Lamé and Clapeyron (1831), Neumann (1860)) of the free boundary problem he posed with measurements obtained from polar explorations in the Canadian arctic. Presumably these data were obtained from the logs of explorers searching for a northwest passage during 1829-1853 (J.C. and H.R. thank C. Vuik for sharing this historical information from his thesis).

Financial support was received from École Polytechnique (Québec), FCAR (Québec), NASA (USA), NSERC (Canada) and Université de Montréal (Québec). We are grateful to M. Glicksman for organizing a wonderful session on microgravity sponsored entirely by NASA (USA). Finally we thank the director, F. Clarke, and the members and staff of the Centre de Recherches Mathématiques (Montréal) for providing the major part of the funding for the meeting and for their kind hospitality and assistance during our stay in Montreal. We shall be forever indebted to S. Chênevert, C. Doré and J. Roy for facilitating every aspect of the organization of the meeting with professional efficiency and charm.

The main theme of this volume is the concept of free boundary problems associated with solids. The first free boundary problem, the freezing of water - the Stefan problem, is the prototype of solidification problems which form the main part of this volume. The two sections treating this subject cover a large variety of topics and procedures, ranging from a theoretical mathematical treatment of solvability to numerical procedures for practical problems. Some new and interesting problems in solid mechanics are discussed in the first section while in the last section the important and new subject of solid-solid phase transition is examined.

Hamilton and London, Canada   J. Chadam
November, 1992   H. Rasmussen

Solid mechanics

S N ANTONTSEV AND S I SHMAREV

Local energy method and vanishing properties of weak solutions of quasilinear parabolic equations

In this work we study the localization (vanishing properties of weak solutions of nonlinear parabolic equations in the form

$$\frac{\partial}{\partial t}(|u|^{\alpha-1}u) = div(\mathbf{A}(x,t,u,\nabla u)) - B(x,t,u) + f(x,t) + div\mathbf{g}(x,t) \qquad (1)$$

$$(x,t) \in Q = \Omega \times (0,T), \ |Omega \in R^N$$

We use a modified method of local energy estimates introduced, justified and widely applied in [1-6]. The modification of local energy method employed enables us to prove, firstly, the validity of some already known effects under weaker assumptions: on the data; and, secondly, to discover new forms of localization, in particular, with $f \not\equiv 0$, $g \not\equiv 0$.

From now on we consider Eq. (1) with the initial condition

$$u(x,o) = u_0(x) \qquad (2)$$

and assume the following assumptions to be fulfilled:

$$\mathbf{A} : M_1|\mathbf{q}|^p \leq (\mathbf{A}(x,t,r,\mathbf{q}),\mathbf{q}) \leq M_2|bfq|^p$$

$$r \in R, \ \mathbf{q} \in R^N \qquad (3)$$

$$\mathbf{B} : B(x,t,r) \geq M_3|r^{\lambda-1}r$$

where $M_i > 0, \alpha > 0, C > 0, \lambda > 0, p > 1$ are given structural constants. We shall consider weak solutions of eq. (1).

$$u(x,t) \in L^p_{loc}(0,T;W^1_p(\Omega)) \cap L^\infty(0,T;L^{\alpha+1}_{loc}(\Omega))$$

Let us introduce (see Fig. 1)

$$B(r, x_0) = \{x \in R^N : |x - x_0| < r\}, \ r > 0, \ x_0 \in |Omega$$

$$K(r, t, x_0, X) = \{(x, \theta) \in Q : |x - x_0| < r + X\theta, \ \theta \in (0, t)\} \qquad (4)$$

$$S(r, t, x_0) = B(r, x_0) \times (0, t)$$

$$K(0, t, x_0, X) = \{(x, \theta) \in Q : |x - x_0| < X\theta, \ \theta \in (0, t)\}$$

and

$$E(r, t, x_0, X) = E(r) = \int \int_{K(r, t, x_0, X)} |\nabla u(x, \theta)|^p \, dx \, d\theta$$

$$C(r) = \int \int_{K(r)} |u(x, \theta)|^{\lambda+1} \, dx \, d\theta, \ K(r, t, x_0, X) = K(r)$$

$$G(r) = \operatorname{ess\,sup}_{\tau \in (0, t)} \int_{B(r + Xt, x_0)} |u(x, t)|^{\alpha+1} \, dx \qquad (5)$$

$$\rho(u) = \int \int_Q \{|\nabla u|^p + |u|^{\lambda+1}\} \, dx \, dt + \operatorname{ess\,sup} \int_\Omega |u(x, \tau)|^{\alpha+1} \, dx$$

I. Vanishing in finite time.

Let $\Omega \in R^N$ be a bounded domain and $u(x, t)$ be a weak solution of eq. (1). Assume

$$u|_{\partial\Omega} = 0 \qquad (6)$$

Theorem 1. (see Fig. 2) Let conditions (3), (6) hold and

$$M_3 = 0, \ N \le p \ \text{or} \ 1 + \alpha \le \frac{Np}{N - p}$$

$$u_0(x) \in L^{(\alpha+1)/\alpha}(\Omega), \ f \in L^\infty(0, T; L^{(\alpha+1)/\alpha}(\Omega))$$

$$g_i \subset L^\infty(0, T; L^{p/(p-1)}(\Omega)), \ i = 1, ..., N$$

Assume $f = 0$ and $\mathbf{g} = 0$ for any $t \ge t_0$. If $p < 1 + \alpha$ the relation

$$\|u_0\|_{\alpha+1, \Omega}^{\alpha+1} + T_0^\sigma \int_0^{t_0} \frac{F(\tau)}{(t_0 - \tau)^\sigma} \, d\sigma \le C \cdot t_0^\sigma$$

3

holds with

$$\sigma = \frac{\alpha+1}{\alpha+-p}, \quad F(t) = \|f(\cdot,t)\|_{(\alpha+1)/\alpha,\Omega}^{p/p-1} + \|g\|_{p/(p-1),\Omega}^{p/p-1}$$

and C some suitable constant, (depending only on Ω, α, P and N) , then

$$u(x,t) \equiv 0 \;\; in \;\; \Omega \;\; for \;\; any \;\; t \gtrless t_0$$

Remark 1 . A similar result is valid for the weak solution of eq. (1) with $p \geq 1 + \alpha$, $M_3 >$) and $0 < \lambda < \alpha$ as well as for the weak solution of the Cauchy problem if $\alpha + 1 = N_p/(N-p)$.

2. Nonpropagation of disturbances from the data.

Theorem 2. ("stable" localization, see Fig. 3). Let the following conditions hold:

$$max(\lambda,\alpha) < p - 1 :$$

$$u_0(x) \equiv 0 \;\; in \;\; B(r,x_0), \;\; f(x,t) \equiv 0, \;\; in \;\; S(r_0,T,x_0), \;\; r_0 >)$$

$$I(r_0,T,x/2) \equiv \int_{r_0}^{r} (\sigma - r_0)^\mu \cdot \left(\frac{\alpha}{\alpha+1}\|u_0\|_{\alpha+1,B(\sigma,x_0)}^{\alpha+1}\right.$$
$$\left. + \chi\|f\|_{(\lambda+1)/\lambda,K(\sigma,T,x_0,\pi/2)}^{(\lambda+1)/\lambda}\right)^\gamma d\sigma \leq q < \infty$$

with some

$$\gamma \in \left(\frac{p}{p-1}\left(\frac{\alpha}{\alpha+1} + \theta\left(\frac{1}{\alpha+1} - \frac{1}{P}\right)\right), 1\right)$$

Here

$$\mu = -1 - \delta/(\delta - 1), \quad \chi = (2/M_3)^\lambda \frac{\lambda}{(1+\lambda)^{(1+\lambda)/\lambda}}$$

$$\delta = -\frac{(p(\alpha+1) + N(p-1-\lambda))(N(p-1-\lambda) + \lambda + 1)}{(\alpha+1)(p-1)(N(p-1-\lambda) + p(\lambda+1))}$$

Then there exists a constant $C_* > 0$, depending only on q, r_0 and structural constants, such that any weak solution of eq. (1), satisfying the condition $\rho(u) \leq C_*$, vanishes in $S(r_0,T,x_0)$

$$u(x,t) \equiv o \;\; in S(r_0,T,x_0)$$

Remark 2. One may consider the case $M_3 > 0$ in a exactly similar way. '

3. Expansion of null-set.
Theorem 3. (see Fig. 4). Assume that

$$\lambda + 1 < 1 + \alpha < p$$

$$\frac{\alpha - \lambda}{\alpha + 1} \geq \frac{N(\alpha - \lambda) + \lambda + 1}{N(\alpha - \lambda) + (\alpha + 1)(\lambda + 1)}(1 - \frac{\lambda + 1}{p})$$

$$u_0(x) \equiv 0 \ in \ B(r_0, x_0), \ r_0 > 0$$

$$f(x, t) \equiv 0 \ in \ K(r_0, T, x_0, X), \ X \in (0, \infty)$$

$I(r_0, T, \nu) \leq q < \infty$ with some γ and $\nu = \arctan X$. Then there exists $C_* = \text{const.}$ > 0, $t_* = \text{const} < T$, depending only on q, r_0 and structural constants, such that the condition $\rho(u) \leq C_*$, implies:

$$u(x, t) \equiv 0 \ in \ K(r_0, t_*, x_0, X)$$

4. Occurrence of null-set.
Theorem 4. (see,Fig. 5). Assume that:

$$1 + \lambda < 1 + \alpha < p, \ \alpha < \frac{p - 1}{p}(1 + \lambda)$$

$$= \int_{\partial K} \mathbf{A} \cdot \mathbf{n} u d\Gamma d\theta + \int_{K(r)} u f dx d\theta$$

$$+ \frac{\alpha}{\alpha + 1} \|u_0\|_{\alpha+1, B(r)}^{\alpha+1} + \frac{\kappa_0 \alpha}{\alpha + 1} \|u\|_{\alpha+1, K(r)}^{\alpha+1} \tag{7}$$

Then, using Young's inequality and (3), (5), we obtain the main inequality

$$\frac{\alpha}{\alpha + 1} G(r) + \frac{M_3}{2} C(r) + M_1 E$$

$$\leq M_2 \|\nabla u\|_{P, \partial K}^P \cdot \|u\|_{P, \partial K} + \frac{\alpha \kappa}{\alpha + 1} \|u_0\|_{\alpha, \partial K}^{\alpha+1} + g(r) \tag{8}$$

$$\equiv I_1 + I_2 + g(r)$$

where

$$g(r) = \frac{\alpha}{\alpha + 1} \|u_0\|_{\alpha+1, B}^{\alpha+1} + (\frac{2}{M_3})^{1/\lambda} \frac{1}{(1 + \lambda)^{(\lambda+1)/\lambda}} \|f\|_{(1+\lambda)/\lambda, K}^{(1+\lambda)/\lambda}$$

Step 2. Obtaining the differential inequality.

5

If $\kappa \neq 0$ and $r_0 = \min r > 0$, (Theorems 3, 4), then we have to obtain, in contrast with the previous works, an estimate for the last term in the right hand side of (8). In Theorems 1-2, $\kappa = 0, r_0 > 0$, and all the estimates are similar to those of [6]. In the Theorem 3 $\kappa > 0$, but $r_0 > 0$, and we may use an embedding theorem in the form

$$\|u\|_{\alpha+1,\partial K}^{\alpha+1} \leq C(r_0, 1/r_0)(\|\nabla u\|_{P,K}^P + \|u\|_{\lambda+1,k}^{\lambda+1})$$

In the assumptions of Theorem 4 a straightforward application of this embedding theorem is unjustified as $r_0 = 0$. In the proof of Theorem 4 we used the following inequalities:

$$I_2 \leq \frac{\kappa\alpha}{\alpha+1}(mes\partial K)^{(1-\alpha+\epsilon)/P} \cdot \left(\frac{dy}{dz}\right)^{(P-1)/P} \cdot \|u\|_{P,\partial K}^{\alpha-\epsilon}$$

$$\|u\|_{P,\partial K} \leq a(y+G)^{\frac{\theta}{P}+\tau\frac{1-\theta}{\alpha+1}}, \quad \tau \in (\frac{1+\alpha}{P}, 1)$$

where $y = E + C$, and a depends on $\kappa, alpha, \lambda, P, q, N, M_1 - M_2, \tau$ and t. Finally, we obtain the ordinary differential inequality for "energy" function $y = t(r)$

$$Gy^\gamma \leq dy' + \Phi, \gamma \in (0,1) \tag{8}$$

where $G = const > 0, d(r,t), \Phi(r,t)$ are known functions. The study of the inequality (8) completes the proof of Theorem 4.

REFERENCES

1. S.N.Antontsev. On localization of solutions of nonlinear degenerate elliptic and parabolic equations. Dokl. Akad. Nauk SSSR 260 (1981), 1289-1293. English transl. in Soviet Math. Dokl. 24 (1981).

2. J.I. Diaz and L. Veron C. R. Akad. Sci. Paris Ser. I Math. 297 (1983), 149-152.

3. S.N. Antontsev. Metastability localization of the solution of degenerate parabolic equations in general form. Dynamics of Continuum Medium, Vol. 83. Novosibirsk, 1987, pp.138-144.

4. S.N. Antontsev and J.I. Diaz. New results of nonlinear elliptic and parabolic equations, obtained via energy methods. Dokl. Akad. Nauk SSSR, Vol. 303 (1988), No. 3, pp.524-528.

5. S.N. Antontsev and J.I. Diaz. Applications of the energy method for localization of the solutions of equations of continuum mechanics. Dokl. Akad. Nauk SSSR, Vol. 303 (1988), No. 2, pp.320-325.

6. S.N. Antontsev and S.I. Shmarev. On localization of solutions of nonlinear parabolic equations with linear sources of general form. Dinamika Sploshnoi Sredy, Vyp. 89 (1989), Novosibirsk, pp.28-42.

S. N. Antontsev, S. I. Shmarev   Lavrentyev Institute of Hydrodynamics
Siberian Division of the USSR Academy of Sciences   Novosibirsk 630090 USSR

D BLANCHARD AND P NICOLAS
The dissipation as a tool to account for trapped internal variables in a convex set

1. Introduction.

Many papers attempt to model the behavior of standard materials in which several variables β are constrained to belong to a closed convex set C (see [1], [4], [5], [6], [7]). To this effect the indicator function I_C of C is added to the usual free energy $\psi(T, \varepsilon, \alpha, \beta)$ (a regular function of the temperature T, the linearized deformation ε and the internal variables α, β).

In this paper, we investigate a different technique which encompasses a more general case. Let C be a closed convex set in the Euclidean space of a few internal variables β; the evolution of the medium must be such that the distance between these variables and C is non-increasing in time. Thus if at some time t_0 the convex constraint is satisfied, i.e. if these variables lie in C at time t_0, then the constraint holds at any time larger than t_0. Furthermore, the variables under consideration are not a priori restricted by any kind of constraint.

We consider a medium whose reversible behavior is described through its free energy $\psi(T, \varepsilon, \alpha, \beta)$. The internal variables β are constrained to the following condition :

(1) the distance between $\beta(t)$ and C is non-increasing in time.

The volumic entropy s, the reversible stress σ^r, the thermodynamical forces A, B respectively related to α, β are given by :

$$(2) \qquad s = - \rho \frac{\partial \psi}{\partial T}, \quad \sigma^r = \rho \frac{\partial \psi}{\partial \varepsilon}, \quad A = - \rho \frac{\partial \psi}{\partial \alpha}, \quad B = - \rho \frac{\partial \psi}{\partial \beta}.$$

Upon decoupling the intrinsic dissipation d_1 from the thermal dissipation and assuming the validity of the Helmholtz postulate, one is led to set, by virtue of Clausius-Duhem inequality,

$$(3) \qquad d_1 = (\sigma - \sigma^r) : \dot{\varepsilon} + A . \dot{\alpha} + B . \dot{\beta} \geqslant 0$$

where σ is the Cauchy stress tensor and overdot denotes time derivative.

For standard media ([8]), a dissipation potential $\Phi(\sigma - \sigma^r, A, B; \varepsilon, \alpha, \beta)$ of the variables $\sigma - \sigma^r$, A and B assumed to be convex, non-negative with zero value at zero in these variables is introduced. Inequality (3) is then insured through the constitutive laws

$$\dot{\varepsilon} \in \partial_{\sigma - \sigma^r} \Phi(\sigma - \sigma^r, A, B; \varepsilon, \alpha, \beta)$$

$$\dot{\alpha} \in \partial_A \Phi(\sigma - \sigma^r, A, B; \varepsilon, \alpha, \beta)$$

$$(4) \qquad \dot{\beta} \in \partial_B \Phi(\sigma - \sigma^r, A, B; \varepsilon, \alpha, \beta)$$

where $\partial_Y \Phi$ denotes the subdifferential of Φ with respect to Y.

The purpose of this paper is to associate to the convex set C a class of dissipation potentials Φ such that law (4) insures that condition (1) be satisfied.

2. Dissipation potential associated to C.

Denote by I_C the indicator function of the closed convex set C and by $\partial I_C(\beta)$ its subdifferential at β. Let $\hat{\beta}$ be the projection of β on C, so that $\partial I_C(\hat{\beta})$ is a closed convex cone with vertex zero. Let $K(\beta)$ be a closed convex cone with vertex zero such that

$$(5) \qquad \beta - \hat{\beta} \in K(\beta) \subset \partial I_C(\hat{\beta})$$

8

and denote by $K^0(\beta)$ its polar cone

$$K^0(\beta) = \{\lambda, \ \lambda.w \leqslant 0 \text{ for any } w \text{ in } K(\beta)\}.$$

Let $\varphi_\beta(\sigma-\sigma^r, A, B)$ be a lower semi-continuous convex function which is non-negative and with zero value at zero; it may exhibit a dependence on β (for the sake of simplicity we omit in this short paper the possible dependences on α and ε). Define the dissipation potential Φ by

$$(6) \qquad \Phi(\sigma - \sigma^r, A, B; \beta) = \inf_{w \in \kappa(\beta)} \varphi_\beta(\sigma - \sigma^r, A, B - w).$$

Remark that the function Φ satisfies the properties that define a dissipation potential. Moreover the following proposition ([2]) is useful in order to derive the constitutive law (4):

<u>Proposition</u> : a) $\partial_B \Phi(\sigma - \sigma^r, A, B; \beta) \subset K^0(\beta)$

b) If $\sigma - \sigma^r$, A and B are such that :

$$(7) \quad \begin{cases} \text{there exists b such that } \partial_B \varphi_\beta(\sigma - \sigma^r, A, B - b) \cap \partial I_{K(\beta)} \ (b) \\ \text{is non empty,} \end{cases}$$

then Φ is subdifferentiable at $(\sigma - \sigma^r, A, B)$ with respect to B and for any b' such that $\partial_B \varphi_\beta(\sigma - \sigma^r, A, B - b') \cap \partial I_{K(\beta)} \ (b')$ is non empty, we have :

$$\partial_B \Phi(\sigma - \sigma^r, A, B; \beta) = \partial_B \varphi_\beta(\sigma - \sigma^r, A, B - b') \cap \partial I_{K(\beta)} \ (b').$$

<u>Remark</u>. The condition (7) is a compatibility condition between φ_β and $K(\beta)$. As an example, it is satisfied when φ_β is continuous at a point of the set $(\sigma - \sigma^r, A, B - K(\beta))$ and the infimum in (6) is achieved.

With the help of a), the constitutive law (4) implies

$$(8) \qquad\qquad\qquad \dot{\beta} \in K^0(\beta),$$

which is essential if the desired behavior (1) is to be achieved. Indeed, under weak regularity assumption on β ([3]), the hypothesis $\beta - \hat{\beta} \in K(\beta)$ together with (8) leads to

$$\frac{1}{2}\frac{d}{dt}\,\|\beta - \hat{\beta}\|^2 = (\beta - \hat{\beta}).(\dot{\beta} - \dot{\hat{\beta}}) = (\beta - \hat{\beta}).\dot{\beta} \leqslant 0$$

because $(\beta - \hat{\beta}).\dot{\hat{\beta}} = \dfrac{dI_C}{dt}\,(\hat{\beta}) = 0$.

Remark. Under the assumption that $\beta(t=0)$ lies in C, the compatibility condition (7) and the constitutive law (4) are equivalent to the usual formulation obtained by resorting to non-smooth free energies as mentioned in the introduction, i.e.

$$B \in \partial I_{K^0(\beta)}(\dot{\beta}) + \partial_{\dot{\beta}}\varphi^*_\beta(\sigma - \sigma_r,\, A,\, \dot{\beta})$$

where φ^*_β is the conjugate of φ_β with respect to B.

3. Applications.

3.1. Phase changes:

Let us consider in this section a two-phase medium and assume that the phase transformation is only due to thermal effects. The medium is described by the temperature T and the volume fraction β of one phase. Setting all the physical parameters equal to one, the free energy is given by

$$\psi(T,\, \beta) = -\,\beta(T - T_0) - T \log T$$

where T_0 is referred as the "phase change temperature". In accordance with the definition of β, we assume that the initial data $\beta(t=0)$ lies between 0 and 1. An illustration of the method developed in section 2 for this straightforward example is provided by setting φ to be equal to $1/2\ (T - T_0)^2$ and $K(\beta)$ is given by

$$K(\beta) = \{w \in \partial I_C(\hat{\beta}),\ w.(\beta - \hat{\beta}) \geqslant 0\}.$$

The compatibility condition (7) is satisfied for any $T - T_0$.

We first examine the reversible case and choose C as the closed interval $[0,1]$. The constitutive law (4) and b) of the foregoing proposition lead to

$$\dot{\beta} = \text{projection}_{K^0(\beta)}\,(T - T_0) = T - T_0 - \text{projection}_{\partial I_C(\beta)}\,(T - T_0)$$

10

which, of course, insures (1) since

$$\dot{\beta} = (T - T_0)^+ \qquad \text{if } \beta = 0,$$

$$\dot{\beta} = - (T - T_0)^- \qquad \text{if } \beta = 1,$$

$$\dot{\beta} = T - T_0 \qquad \text{if } 0 < \beta < 1,$$

where $B^+ = \max(0, B)$, $B^- = \max(0, -B)$ for $B \in \mathbb{R}$.

__Remark__. In our setting the non-dissipative phase change behavior $T - T_0 \in \partial I_C(\beta)$ is seen as a limit of the dissipative law for $\varphi^\varepsilon (T - T_0) = \dfrac{1}{2\varepsilon} (T - T_0)^2$ as ε tends to 0.

Consider now the irreversible case, and let β be the volume fraction of the phase which the medium wants to change to. The variable β will show a trend towards the 1-state, so we choose C as the set $\{1\}$. The constitutive law (4) can be written as

$$\dot{\beta} = (T - T_0)^+ \qquad \text{if } 0 \leqslant \beta < 1,$$

$$\dot{\beta} = 0 \qquad \text{if } \beta = 1.$$

It is a more precise law than that obtained for irreversible phase change in [7]

$$T - T_0 \in \dot{\beta} + \partial I_C(\beta) + \partial I_{\mathbb{R}^+}(\dot{\beta}).$$

3.2. __Damage in an elastic medium__.

An usual technique to model some kind of damage process is to introduce a damage variable D which can be seen as the microporosity in the damage elastic medium ([9]). The variable D is then constrained to lie between 0 and 1. When $D = 0$, the integrity of the medium is preserved whereas $D = 1$ corresponds to a totally damaged situation. A simple example is described by setting

$$\rho \, \psi(\varepsilon, D) = \frac{1}{2} (1 - D) \, E \, \varepsilon : \varepsilon$$

where the density ρ is constant and $E_{ijk\ell}$ is the usual elastic tensor. The dissipation d_1 given by (3) rewrites as

$$d_1 = (\sigma - \rho \, \frac{\partial \psi}{\partial \varepsilon}) : \dot{\varepsilon} + B . \dot{D} \geqslant 0$$

11

with $B = - \rho \dfrac{\partial \psi}{\partial D} = \dfrac{1}{2} E \, \varepsilon : \varepsilon$, the elastic energy of the virgin medium. It is physically reasonable to decouple the two terms in d_1 and thus obtain non-negative $B . \dot{D}$; since B is also non-negative, the damage D is non-decreasing in time. So let us choose $C = \{1\}$ and

$$K(D) = \{w \in \partial I_C(\widehat{D}), \ w.(D - \widehat{D}) \geqslant 0\}.$$

Let us denote by W(D) the threshold of elastic energy B for which the damage occurs ([9]): B must lie in the convex set

$$\mathcal{B}(D) = \{Y, \ Y \leqslant W(D)\}.$$

Taking into account the threshold effect and the elastic behavior (no viscosity), we define

$$\varphi_b (\sigma - \sigma^r, \ B) = I_{\{0\}}(\sigma - \sigma^r) + I_{\mathcal{B}(D)} (B)$$

where $I_{\{0\}}$ and $I_{\mathcal{B}(D)}$ are respectively the indicator functions of $\{0\}$ and $\mathcal{B}(D)$. Upon writing the constitutive laws one gets

$$\sigma = \sigma^r = \dfrac{1}{2} (1 - D) \, E \, \varepsilon$$

$$\dot{D} \in \partial I_{\mathcal{B}(D)} (B) \cap \partial I_{K(D)} (0).$$

The last relation expresses that

$$\dot{D} \geqslant 0 \ ;$$

further,

- if $0 \leqslant D < 1$ * if $B < W(D)$ or if $B = W(D)$ with $\dot{B} - W'(D).\dot{D} < 0$ then $\dot{D} = 0$,

     * if $B = W(D)$ with $\dot{B} - W'(D).\dot{D} \geqslant 0$ then $W'(D).\dot{D} = A\varepsilon : \dot{\varepsilon}$,

- if $D = 1$ then $\dot{D} = 0$.

The above obtained model permits to recover the usual laws for such problems (c.f. e.g. [9] pp. 430-431).

12

REFERENCES

[1] D. BLANCHARD, *Etude de problèmes d'évolution en mécanique des milieux dissipatifs*, Thèse d'Etat, Université Pierre et Marie Curie, Paris, 1986.

[2] D. BLANCHARD, P. NICOLAS, *Dissipation dans les matériaux à variables internes piégées dans un convexe*, C.R. Acad. Sc. Paris, to appear.

[3] H. BREZIS, *Opérateurs maximaux monotones et semigroupes de contraction dans les espaces de Hilbert*, North-Holland, Amsterdam, 1973.

[4] M. FREMOND, *Matériaux à mémoire de forme*, C.R. Acad. Sc. Paris, t. 304, Série II, n°7, 1987, p. 239-244.

[5] M. FREMOND et P. NICOLAS, *Hystérésis dans les milieux poreux humides non saturés*, C.R. Acad. Sc. Paris, t. 305, Série II, 1987, p. 741-746.

[6] M. FREMOND et P. NICOLAS, *Macroscopic thermodynamics of porous media*, Continuum Mech. and Thermodyn., Vol. 4, 1990, Springer-Verlag, to appear.

[7] M. FREMOND et A. VISINTIN, *Dissipation dans le changement de phase. Surfusion. Changement de phase irréversible.* C.R. Acad. Sc. Paris, t. 301, Série II, 1985, p. 1265-1268.

[8] B. HALPHEN et NGUYEN QUOC SON, *Sur les matériaux standard généralisés*, J. de Mécanique, 14, 1975, p. 39-63.

[9] J. LEMAITRE et J.L. CHABOCHE, *Mécanique des matériaux solides*, Dunod, Paris, 2ème édition, 1988.

Dominique BLANCHARD et Pierre NICOLAS
Service de Mathématiques
Laboratoire Central des Ponts et Chaussées
U.M.R. 113 LCPC/CNRS
58 Boulevard Lefebvre
75732 PARIS CEDEX 15

P COLLI AND A VISINTIN
Doubly nonlinear evolution equations accounting for dissipations

1. Introduction. Under the assumption of *normal dissipativity* (see [18]), partial differential equations of the following type

$$(1) \qquad \alpha\left(\frac{\partial u}{\partial t}\right) - \operatorname{div}\left(\beta(\nabla u)\right) \ni f,$$

where α and β are maximal monotone graphs (and possibly subdifferentials of proper convex lower semicontinuous functions), model the behaviour of a class of elastic viscoplastic and elastic plastic materials (cf. [16, 18]). In general the graphs α and β are multivalued so that (1) yields a class of *possibly degenerate parabolic equations*. It is not difficult to see that some (double) variational inequalities can be represented in the form (1) and free boundary problems are considered, according to the choice of α and β.

Here, following [11], we give existence results for the initial value problem corresponding to (1) in an abstract setting (which includes also other applications). Namely we consider the equation

$$(2) \qquad A\frac{du}{dt} + Bu \ni f$$

for A and B maximal monotone operators in a Hilbert space H, with A bounded, B unbounded and such that its domain $D(B)$ is contained in a Banach space V embedded compactly in H. One requires that at least one of the two operators be the subdifferential of a convex and lower semicontinuous function and assumes suitable coerciveness conditions. Arguments are based on monotonicity and compactness techniques.

It does not seem that equation (2) has yet been studied, unless one of the operators is linear. In fact, the case with A linear (and self-adjoint) is well known and has been first analysed by Brezis [9] (see also [4], [21],

14

and [10] for related literature), while equations with B linear arise from heat control problems, e.g., and have been studied by Duvaut and Lions [13,14]. Otherwise, the authors just know of an equation of the form $\alpha(\partial u/\partial t) + \beta(u) \ni f$, coupled with a degenerate diffusion equation, studied by Blanchard, Damlamian, and Guidouche [7]. More concern has been devoted to equations of the form

(3)
$$\frac{d}{dt}(Au) + Bu = f \qquad \text{in } H,$$

still with A and B nonlinear and B unbounded in H (see, e.g., [20, 17,2,5,3,12,1,6,8]). Hovever one can easily see that equations (2) and (3) are different and cannot be represented in the same way.

2. Applications. In this section we will illustrate our existence results for the initial value problem related to (2) on an example dealing with equation (1) in Sobolev spaces. Also, we point out some applications.

Let $\Omega \subset \mathbf{R}^M$ $(M \geq 1)$ be a bounded domain and let α, β be two maximal monotone graphs of $\mathbf{R}^N \times \mathbf{R}^N$ $(N \geq 1)$ and of $\mathbf{R}^{M \times N} \times \mathbf{R}^{M \times N}$, respectively. Given a function $f : Q := \Omega \times]0, T[\to \mathbf{R}^N$, we look for a function $u : Q \to \mathbf{R}^N$ satisfying (1) and suitable initial and boundary conditions. More precisely, (1) has to be understood as equivalent to the system

(4) $w + v = f, \quad w \in \alpha \left(\dfrac{\partial u}{\partial t} \right), \quad v \in - \operatorname{div} (\beta(\nabla u)) \quad \text{in } \mathcal{D}'(Q).$

We can prove existence of a solution of the corresponding initial-boundary value problem, uniqueness being an open question if both α and β are nonlinear.

First we require that α be coercive and with linear growth at infinity, and β the subdifferential of a proper convex lower semicontinuous function $\psi : \mathbf{R}^{M \times N} \to \mathbf{R}$ such that for some $a > 0$, $b \in \mathbf{R}$,

$$\psi(v) \geq a|v|^2 - b \qquad \forall v \in \mathbf{R}^{M \times N}.$$

This setting corresponds to a *nonlinear relaxation dynamics* for a potential system. Here one can consider an approximate equation (depending on a parameter $\varepsilon > 0$) and multiply it by $(\partial u_\varepsilon / \partial t)$; this yields the estimate

(5) u_ε is uniformly bounded in $H^1(0, T; L^2(\Omega)^N) \cap L^\infty(0, T; H^1(\Omega)^N).$

15

Hence, possibly extracting a subsequence, u_ϵ weakly star converges to some u in the above space. By means of a standard monotonicity and compactness procedure, one can then show that

$$(6) \qquad v := -\lim_{\epsilon \searrow 0} \text{div} \left(\beta_\epsilon(\nabla u_\epsilon) \right) \in -\text{div} \left(\beta(\nabla u) \right) \quad \text{in } \mathcal{D}'(Q).$$

Finally, as β is *cyclically monotone* (being a subdifferential [9]), by using (1) and (5), one can prove that

$$(7) \qquad w := \lim_{\epsilon \searrow 0} \alpha_\epsilon \left(\frac{\partial u_\epsilon}{\partial t} \right) \in \alpha \left(\frac{\partial u}{\partial t} \right) \quad \text{in } \mathcal{D}'(Q).$$

Our other existence results require α to be linearly bounded and equal to the subdifferential of a proper convex lower semicontinuous function φ : $\mathbf{R}^N \to \mathbf{R}$ (without any coerciveness assumption), and β to be strongly monotone and either Lipschitz continuous or equal to the subdifferential of a proper convex lower semicontinuous function ψ : $\mathbf{R}^{M \times N} \to \mathbf{R}$. In this setting the function φ can correspond to a *dissipation potential*. Also here a convenient approximation is introduced, then the equation is differentiated in time and multiplied by $(\partial u_\epsilon / \partial t)$; this yields the estimate

$$(8) \qquad u_\epsilon \text{ is uniformly bounded in } H^1(0, T; H^1(\Omega)^N),$$

which allows us to take the limit in the approximate equation without much difficulty.

Equations of the form (2) occur in several physical models. For instance in thermodynamics, denoting by u the vector of *generalized displacements* and by F that of *generalized forces*, from the *second principle of thermodynamics* it follows that the so-called *phenomenological laws* are of the form

$$(9) \qquad F = Bu,$$

with B monotone and $B0 = 0$ [16]. In a neighbourhood of $u = 0$ one can assume that B is linear; moreover, by *Onsager relations*, B is also self-adjoint, hence *cyclically monotone*, that is $B = \partial \psi$, with ψ convex potential. We allow ψ to be nonquadratic. Now one can introduce the assumption of *normal dissipativity* [18], requiring the existence of another convex function φ, named *dissipation potential*, such that

$$(10) \qquad \partial \varphi \left(\frac{du}{dt} \right) = -F.$$

16

Thus by (9) and (10) we have

(11)
$$\partial\varphi\left(\frac{du}{dt}\right) + \partial\psi(u) \ni 0.$$

More generally the presence of an exterior thermodynamic force $-f$ would yield a right hand side f.

3. Results.

In this section we state precisely the results announced above. Let H be a real Hilbert space which we identify with its dual, V a reflexive Banach space dense and compactly embedded in H, and V' the dual space of V. We denote by $(\cdot\,,\cdot)$ either the scalar product in H or the duality pairing between V' and V, and by $|\cdot|$ the norm in H. Besides, let A and B be maximal monotone operators in H with domains $D(A)$ and $D(B)$. The operator A is assumed to be bounded in H, so that A^{-1} is surjective and $D(A) \equiv H$ (see [9]), while B is unbounded and such that $D(B) \subset V$.

Theorem 1. *Assume that*

 (i) $\exists\, C_1 > 0 : \forall\, u \in H, \forall\, \xi \in Au \qquad (\xi, u) \geq C_1\left(|u|^2 - 1\right),$

 (ii) $\exists\, C_2 > 0 : \forall\, u \in H, \forall\, \xi \in Au \qquad |\xi| \leq C_2\left(|u| + 1\right),$

 (iii) *B is the subdifferential of a proper, convex and lower semicontinuous function* $\psi : H \to\,]-\infty, +\infty],$

 (iv)   *$D(\psi) \subset V$ and there exist $C_3, C_4 > 0$, $p_1, p_2 > 0$ such that*
$$\psi(u) \geq C_3\, \|u\|_V^{p_1} - C_4\left(|u| + 1\right)^{p_2} \qquad \forall\, u \in V,$$

 (v) $f \in L^2(0, T; H),\quad u_0 \in D(\psi).$

Then there exist $u \in H^1(0, T; H) \cap L^\infty(0, T; V)$ and $w, v \in L^2(0, T; H)$, such that, setting $u' = (du/dt)$, one has

(12) $w(t) + v(t) = f(t),\; w(t) \in Au'(t),\; v(t) \in Bu(t)$ *for a.e. $t \in\,]0, T[$,*

(13)
$$u(0) = u_0.$$

Remark 1. Assumptions (i)-(ii) restrict the behaviour of A at infinity, but allow for the presence of *horizontal segments* in this graph.

Theorem 2. *Assume that*

 (vi)   *A is the subdifferential of a proper, convex and lower semicontinuous function* $\varphi : H \to\,]-\infty, +\infty],$

17

(vii) $B : V \to V'$ *is a Lipschitz continuous operator,*

(viii) B *is strongly monotone, i.e. there is* $C_5 > 0$ *such that*
$$\forall\, u_1, u_2 \in V \quad (Bu_1 - Bu_2, u_1 - u_2) \geq C_5 \|u_1 - u_2\|_V^2\,,$$

(ix) $f \in H^1(0, T; V')$, $\quad u_0 \in V$, $\quad f(0) - Bu_0 \in D(\varphi^*)$,

where φ^* *is the convex conjugate function of* φ. *Then there exist* $u \in H^1(0, T; V)$ *and* $w \in L^\infty(0, T; H) \cap H^1(0, T; V')$ *satisfying (13) and such that for a.e.* $t \in]0, T[$

(14) $w(t) + Bu(t) = f(t) \quad in \ V', \quad w(t) \in Au'(t).$

Remark 2. It is easy to see (cf., e.g., [**19**]), that by (vii) and (viii) the restriction of the operator B to $D(B) \subset V$ taking values in H is maximal monotone in H and surjective: for any $g \in H$ the equation $Bu = g$ has one and only one solution $u \in V$.

Remark 3. Let B satisfy (iii) and let $D(\psi) \subset V$, so that ψ is proper, convex and lower semicontinuous also in V. Denote by $\widetilde{B}$ the subdifferential of ψ restricted to V and by $\widetilde{D} \subset V$ its domain. Obviously $\widetilde{B} : \widetilde{D} \to V'$ is an extension of B and $D(B) \subset \widetilde{D} \subset D(\psi)$.

Theorem 3. *Let* $D(\psi) \subset V$ *and let the operator* $\widetilde{B}$ *of Remark 3 be strongly monotone in* $\widetilde{D}$ *in the sense of assumption (viii). Assume that* $u_0 \in D(B)$ *and that (vi), (iii) and (ix) (with* Bu_0 *replaced by the minimal norm element* $B^0 u_0$*) hold. Then there exist* $u \in H^1(0, T; V)$, $w \in L^\infty(0, T; H)$, $v \in L^\infty(0, T; V')$, *satisfying (13) and such that for a.e.* $t \in]0, T[$

(15) $w(t) + v(t) = f(t), \quad v(t) \in \widetilde{B}u(t) \quad in \ V', \quad w(t) \in Au'(t).$

Remark 4. By comparison in the corresponding equations, it is straightforward to see that in Theorem 2 (resp. 3) if $f \in L^2(0, T; H)$, then $Bu \in L^2(0, T; H)$ (resp. $v \in L^2(0, T; H)$), and all equations hold in H.

The proof of the three existence theorems is obtained via a regularization – a priori estimate – passage to the limit procedure using monotonicity and compactness techniques (for more details we refer to [**11**]).

Remark 5. It is not difficult to show that there is at most one solution of problem (12)-(13) if at least one of A or B is *linear* and *self-adjoint*, and moreover at least one of these operators is *strictly monotone* in H. Uniqueness remains an open question if both operators are nonlinear.

18

REFERENCES

[1] H. W. Alt and S. Luckhaus, *Quasilinear elliptic-parabolic differential equations*, Math. Z. **183** (1983), pp. 311-341.

[2] A. Bamberger, *Étude d'une équation doublement non linéaire*, Internal Report, École Polytechnique de Palaiseau, France, 1975.

[3] V. Barbu, *Existence for non-linear Volterra equations in Hilbert spaces*, SIAM J. Math. Anal. **10** (1979), pp. 552-569.

[4] C. Bardos and H. Brézis, *Sur une classe de problèmes d'évolution non linéaires*, J. Differential Equations **6** (1969), 345-394.

[5] P. Benilan, *Sur un problème d'évolution non monotone dans $L^2(\Omega)$*, Internal Report, Université de Besançon, France, 1975.

[6] F. Bernis, *Existence results for doubly nonlinear higher order parabolic equations on unbounded domains*, Math. Ann. **279** (1988), pp. 373-394.

[7] D. Blanchard, A. Damlamian and H. Guidouche, *A nonlinear system for phase change with dissipation*, to appear on Differential Integral Equations.

[8] D. Blanchard and G. A. Francfort, *Study of a doubly nonlinear heat equation with no growth assumptions on the parabolic term*, SIAM J. Math. Anal. **19** (1988), pp. 1032-1056.

[9] H. Brézis, *Opérateurs maximaux monotones et semi-groupes de contractions dans les espaces de Hilbert*, North-Holland, Amsterdam, 1973.

[10] R. W. Carroll and R. E. Showalter, *Singular and degenerate Cauchy problems*, Mathematics in Science and Engineering, Vol. 27, Academic Press, New York, 1976.

[11] P. Colli and A. Visintin, *On a class of doubly nonlinear evolution equations*, to appear on Comm. Partial Differential Equations.

[12] E. Di Benedetto and R. E. Showalter, *Implicit degenerate evolution equations and applications*, SIAM J. Math. Anal. **12** (1981), pp. 731-751.

[13] G. Duvaut and J. L. Lions, *Sur des nouveaux problèmes d'inéquations variationelles posés par la Mécanique. Le cas d'évolution*, C. R. Acad. Sci. Paris **269** (1969), pp. 570-572.

[14] G. Duvaut and J. L. Lions, *Inequalities in mechanics and physics*, Springer-Verlag, Berlin, 1976.

[15] I. Ekeland and R. Temam, *Analyse convexe and problèmes variationnels*, Dunod Gauthier-Villars, Paris, 1974.

[16] P. Germain, *Cours de mécanique des milieux continus*, Masson, Paris, 1973.

[17] O. Grange and F. Mignot, *Sur la résolution d'une équation et d'une inéquation paraboliques non linéaires*, J. Funct. Anal. **11** (1972), pp. 77-92.

[18] B. Halpen and Q. S. Nguyen, *Sur les matériaux standard généralisés*, J. Mécanique **14** (1975), pp. 39-63.

[19] J. L. Lions, *Quelques méthodes de résolution des problèmes aux limites non linéaires*, Dunod Gauthier-Villars, Paris, 1969.

[20] P. A. Raviart, *Sur la résolution de certaines équations paraboliques non linéaires*, J. Funct. Anal. **5** (1970), pp. 299-328.

[21] R. E. Showalter, *Nonlinear degenerate evolution equations and partial differential equations of mixed type*, SIAM J. Math. Anal. **6** (1975), pp. 25-42.

PIERLUIGI COLLI
Dipartimento di Matematica
Università di Pavia
Strada Nuova, 65
27100 Pavia, Italy

AUGUSTO VISINTIN
Dip. di Matematica dell'Università
38050 Povo (TN), Italy
and Istituto di Analisi Numerica
del C.N.R., Pavia, Italy

M FRÉMOND

On the flattening of materials

1.   Introduction

When a material flattens it goes from a geometrical dimension (3, 2 or 1) to a lower one.   The properties of the state change are taken into account by the free energy.   There exists a reaction to the flattening, the properties of which are given. The generalized interior forces in the flattened parts satisfy classical equations, for instance the beam equations when flattening from dimension 2 to dimension 1.

Let us consider the movement of a deformable solid whose position at time $t = 0$ is Ω_a.   A material point which is at point $a = (a_\alpha)$ at $t = 0$ is at point $x = \phi(a, t) = (x_i)$ at time t.   We wish to investigate how parts of this solid flatten. It is classical to describe the deformations by the deformation matrix $F = (F_{i\alpha}) = (\frac{\partial x_i}{\partial a_\alpha})$ which can be written as the product of a rotation matrix R and a right stretch matrix W [1], [3].   It can be proved:

Proposition:

There exists one and only one symmetric, semi definite, positive matrix W such that $F = RW$.

If the material cannot flatten the eigenvalues of W, $\lambda_i(W)$, are strictly positive.   If the material can flatten one or several $\lambda_i(W)$ are zero.   Because interpenetration cannot occur, the matrix W satisfies

$$\lambda_i(W) \geq 0, \ i = 1, 2 \text{ and } 3.$$

This condition is equivalent to

(1) $\det W \geq 0$, $\operatorname{tr} \{\operatorname{cof} W\} \geq 0$, $\operatorname{tr} W \geq 0$,

where cof W is the matrix of the cofactors of W.

The states of the material which are physically admissible satisfy (1).   We denote by C the set for these admissible states:

$$C = \{U| \ U \ \epsilon \ S| \det U \geq 0, \ \operatorname{tr} \{\operatorname{cof} U\} \geq 0; \ \operatorname{tr} U \geq 0\}$$

where S is the linear space of the 3 × 3 symmetric matrices with scalar product

$W : N = W_{i\alpha} N_{i\alpha}$.

It can be proved:

Theorem 1. The set C of the admissible states is a convex, closed cone with apex at the origin.

The subdifferential of the indicator function [2] I_c of C ($I_c(U) = 0$ if $U \in C$, $I_c(U) = +\infty$ if $U \in C$) is

$$\partial I_c(U) = \{0\},$$

if $\det U > 0$, tr $\{$cof $U\} > 0$ and tr $U > 0$;

$\partial I_c(U) = \{N | N \in S;$ Na $= 0$ for any eigenvector a of W with eigenvalue strictly positive; $\det N = 0$, tr $\{$cof $N\} = 0$, tr $N \leq 0\}$,

if $\det U = 0$, tr $\{$cof $U\} > 0$ and tr $U > 0$;

$\partial I_c(U) = \{N | N \in S;$ Na $= 0$ for the eigenvector a of W with eigenvalue strictly positive; $\det N = 0$; tr $\{$cofN$\} \geq 0$, tr $N \geq 0\}$,

if $\det U = 0$, tr$\{$cof $U\} = 0$, tr $U \geq 0$;

$\partial I_c(U) = \{N | N \in S | \det N \leq 0;$ tr $\{$cofN$\} \geq 0;$ tr $N \leq 0\}$, if $U = 0$.

2. Free energy.

The free energy depends on the state of the material. This state is specified by the right stretch matrix W. Thus we assume the free energy to depend on W. We do think that the free energy takes into account the whole physical properties of the material. Thus it takes into account the internal constraint $W \in C$. We choose as free energy

$$\Psi_w(W) = \Psi(W) + I_c(W),$$

where Ψ is a smooth function.

Let us note that the free energy Ψ_w is defined for any $W \in \mathbb{R}^6$ and its value is $+\infty$ for any deformation W which is not a possible one. Thanks to proposition 1 and [3] it can be proved that Ψ_w is frame indifferent.

3. Constitutive law.

A constitutive law is a relation between Π (the Boussinesq matrix), F and $\frac{dF}{dt}$ which must satisfy the Clausius–Duhem inequlaity

(2) $\frac{d\Psi}{dt} \leq \Pi : \frac{dF}{dt}$, for any actual evolution.

To get the constitutive laws, we choose first the state laws

(3) $\Pi^e = \frac{\partial \Psi}{\partial F} = R \frac{\partial \Psi}{\partial W}$, $\Pi^r = R \partial I_c(W)$, $(F = RW)$.

The stress Π^e is the elastic or reversible stress. The stress Π^r is the

21

thermodynamical reaction to the internal constraint (1) (or $W \in C$). It is 0 if W is interior to C and it is a stress normal to C if W is on the boundary of C(theorem 1).

One can remark that the reaction Π^r works with a virtual velocity $E \in \mathbb{R}^6$ whereas it does not work with an actual velocity, $\Pi^r: \dfrac{dF}{dt} = 0$.

We assume the difference $\Pi - (\Pi^e + \Pi^r)$ to depend on the rate $\dfrac{dF}{dt}$ and we state the following inequality to be satisfied

(4) $\forall E \in \mathbb{R}^6,\ (\Pi - \Pi^e - \Pi^r): E \geq 0.$

It can be shown that this Clausius–Duhem like inequality (4) extended to virtual rates E and the state laws (3) imply the classical properties which are expected:

<u>Theorem 2.</u> If the relation (3) and (4) are satisfied, then

  i) the internal constraint $W \in C$ is satisfied,

  ii) the Clausius–Duhem inequality (3) is satisfied.

Let us remark that in this setting the fact that $W \in C$ is part of the constitutive laws through the state law, $\Pi^r \in R\ \partial I_c(W)$ because "$\partial I_c(W)$ is not empty" is equivalent to $W \in C$.

Let us also note that the thermodynamical reaction is an uniaxial matrix normal to the flattened shape if the solid flattens from dimension 3 to dimension 2, it is a biaxial matrix normal to the flattened shape if the solid flattens from dimension 3 to dimension 1. It is a complete matrix if the solid flattens to a point.

4. <u>A one dimension example.</u>

Let there be a spring lying in the segment $\Omega_a = \,]0,\ \ell]$. It is fixed at the point 0 and loaded by an increasing compression $f(t)$ ($f(0) = 0$, $f(t) < 0$) which intends to flatten it. By neglecting the inertial forces, we have

$$\frac{\partial \Pi}{\partial a} = 0 \text{ in }]0,\ \ell[\text{ and } \Pi(\ell, t) = 0.$$

We have $W = \dfrac{\partial x}{\partial a}$ and $C = \{W|\ w \geq 0\}$.   We choose as free energy $\Psi(W) = \dfrac{k}{2}\,(W - 1)^2$ with $k > 0$. Then we have

$$\Pi^e = k(W - 1) \text{ and } \Pi^r = -\,p \in \partial I_c(W).$$

The last equation means that the pressure p is zero if $W > 0$ and is positive if $W = 0$. We assume the spring to be elastic: $\Pi = \Pi^e + \Pi^r$. The solution of these equations is– for a weak compression $(- k \le f(t) \le 0)$ the spring is not flattened:

$$x(a, t) = a \left(\frac{f(t) + k}{k}\right), \ k = 0;$$

for a strong compression $(f(t) \le - k)$, the spring is flattened at the point 0,

$$x(a, t) = 0 \text{ and } p = - (f(t) + k) \ge 0.$$

Let us note that the pressure p, the thermodynamical reaction is not equal to the mechanical reaction at the point 0. The thermodynamical reaction depends on the constitutive law (through k) whereas the mechanical reaction equal to –f(t) does not.

5. <u>A three dimension example.</u>

Let there be a cube $\Omega_a = (]0, \ell)^3$. It is loaded on its opposite faces by increasing surfacic compressions in the following order $(f_1(t) \le f_2(t) \le f_3(t), \ f_i(0) = 0,$ $f_i(t) \le 0$. We choose as free energy $\Psi(W) = \frac{k}{2}(W - \hat{I})^2$ where $\hat{I}$ is the unit matrix. By neglecting the inertial forces, the balance equations are

$\Pi_{i\alpha,\alpha} = 0$ in Ω_a and $\Pi_{i\alpha} N_\alpha = 0$ on each face of the cube with unit outwards normal vector N. It is easy to check that the balance equations are satisfied by

$$\Pi = \begin{bmatrix} f_1(t) & 0 & 0 \\ 0 & f_2(t) & 0 \\ 0 & 0 & f_3(t) \end{bmatrix}.$$

Assuming $R = \hat{I}$ the state equations are

$$\Pi^e = k(W - \hat{I}) \text{ and } \Pi^r \ \varepsilon \ \partial I_c(W).$$

We assume the material to be elastic $\Pi = \Pi^e + \Pi^r$. By solving these equations we have,

for a very small compression $(- k \le f_1(t) \le 0)$, the cube is not flattened

$$x_1(a, t) = a_1\left(\frac{f_1(t) + k}{k}\right), \ x_2(a, t) = a_2\left(\frac{f_2(t) + k}{k}\right),$$

$$x_3(a, t) = a_3\left(\frac{f_3(t) + k}{k}\right),$$

and $\Pi^r = 0 \ \varepsilon \ \partial I_c(W)$;

for a small compression $(f_1(t) \le - k \le f_2(t))$. The cube flattens on the faces where the compression f_1 is applied and it becomes a rectangle,

$$x_1(a, t) = 0, \quad x_2(a, t) = a_2\left(\frac{f_2(t) + k}{k}\right), \quad x_3(a,t) = a_3\left(\frac{f_3(t) + k}{k}\right)$$

and $\Pi_r = \begin{vmatrix} f_1(t) + k & 0 & 0 \\ 0 & 0 & 0 \\ 0 & 0 & 0 \end{vmatrix} \varepsilon \, \partial\mathcal{I}_c(W).$

One can check that Π^r is normal to the rectangle;

for a large compression ($f_2(t) \leq -k \leq f_3(t)$) the rectangle flattens into a segment,

$$x_1(a,t) = 0, \quad x_2(a,t) = 0, \quad x_3(a,t) = a_3\left(\frac{f_3(t) + k}{k}\right),$$

and $\Pi^r = \begin{bmatrix} f_1(t) + k & 0 & 0 \\ 0 & f_2(t) + k & 0 \\ 0 & 0 & 0 \end{bmatrix} \varepsilon \, \partial\mathcal{I}_c(W).$

One can check that Π^r is normal to the segment;

for a very large compression ($f_3(t) \leq -k$), the segment flattens into a point,

$$x(a,t) = 0$$

and $\Pi^r = \begin{bmatrix} f_1(t) + k & 0 & 0 \\ 0 & f_2(t) + k & 0 \\ 0 & 0 & f_3(t) + k \end{bmatrix} \varepsilon \, \partial\mathcal{I}_c(0).$

6. The flattened part.

To be simple we assume a part $S_a \subset \Omega_a \subset \mathbb{R}^2$ to flatten into a one dimension line $S_x = \phi(S_a)$. All the points which are on the line of Ω_a

$$\Gamma(x(t)) = \{a \, \varepsilon \, \Omega_a | \ x(t) = \phi(a,t)\}$$

are accumulated at the point $x(t)$ (figure 1). We give briefly the balance laws on the

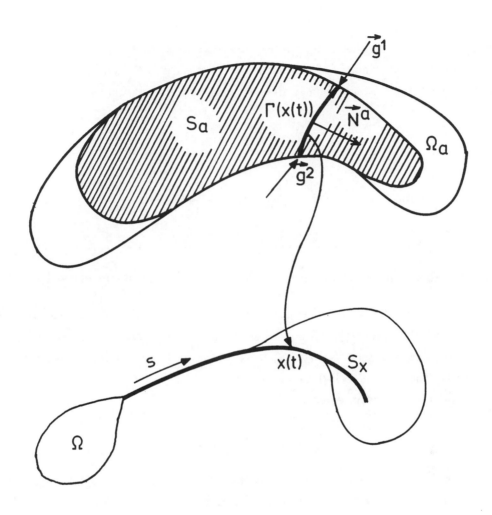

Figure 1

The shaded part S_a is flattened onto the S_x part (thick line) of Ω. The dimension of S_x is one. All the points of the line $\Gamma(x(t))$ are accumulated at the point $x(t)$. The surfacic forces $\vec{g}^1$ and $\vec{g}^2$ are applied to S_a at the ends of Γ by the exterior of S_a.

flattened part. By writing that the mass which is in a slice of S_a is accumulated on a segment of S_x we get the mass balance

$$\rho_x(x,t) = \int_\Gamma \frac{\rho_a(a)}{\text{tr }\{W\}} \, d\gamma$$

where ρ_x and ρ_a are the lineic and surfacic densities in S_x and S_a.

By writing the principle of virtual power for a slice of S_a with virtual velocities compatible with the flattening constraint, we get the rod equation

$$\frac{d\vec{R}}{ds} + \vec{f} = \rho_x \, \vec{A},$$

with

$$R_i = \int_\Gamma \Pi_{i\alpha} N^a_\alpha \, d\gamma, \ f_i = \int_\Gamma \frac{f^a_i}{\text{tr}\{W\}} \, d\Gamma + \frac{g^1_i}{\text{tr}\{W_1\}} + \frac{g^2_i}{\text{tr}\{W_2\}},$$

$$\rho_x \, A_i = \int_\Gamma \rho_a \, A_i \, d\gamma,$$

where $\vec{f}^a$ is the surfacic exterior force in S_a, $\vec{g}^1$ and $\vec{g}^2$ the lineic forces applied by the exterior of S_a to S_a at the ends of Γ. The quantity $\vec{A}$ is the weighted acceleration at the point $x(t)$. There is not a unique acceleration at the point $x(t)$ because there can be a flux of particles in S_x. The quantity s is the curviline absciss on S_x and N_a the normal to Γ directed accordingly to s.

We get also the beam equation [1]

$$\frac{d\vec{M}}{ds} + \vec{T}_x \wedge \vec{R} + \vec{m} = 0,$$

where $\vec{T}_x$ is the tangent vector to S_x. The exterior torque $\vec{m}$ depends on Π but with reasonable assumption this dependence can be removed.

Note. As seen in the examples, if one wishes to know the evolution of a structure, one solves the equations in Ω_a. To the mentioned equations one must add a global non–interpenetration condition and deal carefully with the acceleration because the velocities can be discontinuous [4].

REFERENCES

[1]   P. Germain, Mécanique des milieux continus, Masson, Paris, 1973.

[2]   J. J. Moreau, Fonctionnelles convexes, Séminaire sur les équations aux dérivées partielles, Collége de France, Paris, 1966.

[3]   J. J. Moreau, Lois d'élasticité en grande déformation, Séminaire d'Analyse convexe, Exposé n° 12, Montpellier, 1979.

[4]   J. J. Moreau, An expression of classical dynamics, preprint n°88–1, Laboratoire de Mécanique Générale des Milieux continus, Montpellier, 1988.

Michel Frémond
Laboratoire Mixte LCPC/CNRS: Laboratoire: Laboratoire
des matériaux et des structures
Laboratoire Central des Ponts et Chaussées
58 Boulevard Lefebvre 75732 PARIS CEDEX 15

J M GREENBERG
The long time behavior of elastic-perfectly plastic materials[1]

1. Introduction

My goal in this paper is to present a new modeling approach to elastic-perfectly plastic materials. I shall confine my attention to zero and one dimensional models of such materials. The former may be used to analyze the vibrations of a discrete mass attached to an elastoplastic element and the latter to analyze shear flows in elastoplastic solids. The model used here was developed in [1] and used there to analyze certain signalling problems. Here I shall confine my attention to initial-boundary value problems where the system has no external energy sources. My goal is a description of the long time behavior of these systems and my basic result is that the solutions converge to a state of permanent plastic deformation and ultimately evolve elastically with a state of stress everywhere at or below yield.

I consider the low dimensional models because they avoid geometrical complications and make clear the nature of the yield condition and plastic flow rule. This work is similar to some recent work of Antman and Szymczak [2, 3]. Their model accounts for "strain-hardening", a feature not present here, and also treats the local yield stress as an inviolable constraint. This last point is an essential difference between their models and the ones presented here. The models considered here allow for stresses above yield, but, when this condition obtains, the flow rule forces the plastic strain to change so as to lower the stress levels in the material and dissipate energy. Other recent efforts on the modeling of elastoplastic materials may be found in Coleman and Owen [4], Buhite and Owen [5], Coleman and Hodgdon [6], and Owen [7].

2. Low Dimensional Models

In the first model I assume a mass m connected to linearly elastic spring which in turn is connected to a long weightless rod which fits into a sleeve whose end is located at the origin $x = 0$. The rigid rod is allowed to slip relative the sleeve if the force in the rod is sufficiently large. Figure 1 shows the equilibrium configuration and Figure 2 a typical displaced configuration of the system.

---

[1]This research was partially supported by U.S. National Science Foundation and the U.S. Department of Energy.

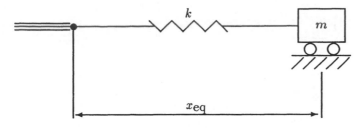

Figure 1

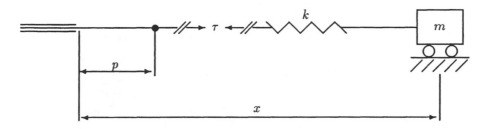

Figure 2

The displacement $\delta(t)$ of the mass m is given by

$$\delta(t) = x(t) - x_{eq}, \qquad (2.1)$$

and the number $p(t)$ represents the amount the rigid rod has slipped at time t. The elastic elongation $e(t)$ of the spring is then given by

$$e(t) = \delta(t) - p(t). \qquad (2.2)$$

The equations of motion for this system are

$$\frac{d\delta}{dt} = v \quad \text{and} \quad m\frac{dv}{dt} = -\tau \qquad (2.3)$$

where of course $v(t)$ is the velocity of the mass m and τ is the restoring force in the spring which is also equal the shear resistance provided by the sleeve on the rigid rod. I assume that

$$\tau = ke, \quad k > 0. \qquad (2.4)$$

To close the system a flow rule is required which describes how the rod slips through the sleeve. I assume the existence of a monotone nondecreasing function $\tau \to s(\tau)$ on $\tau \geq \tau_{\text{yield}}$ and postulate that

$$\frac{dp}{dt} = \begin{cases} s(\tau), & \tau \geq \tau_{\text{yield}} \\ s(\tau_{\text{yield}}), & \tau = \tau_{\text{yield}} \text{ and } v > s(\tau_{\text{yield}}) \\ v, & \tau = \tau_{\text{yield}} \text{ and } 0 \leq v \leq s(\tau_{\text{yield}}) \\ 0, & \tau = \tau_{\text{yield}} \text{ and } v < 0 \\ 0, & -\tau_{\text{yield}} < \tau < \tau_{\text{yield}} \\ 0, & \tau = -\tau_{\text{yield}} \text{ and } v > 0 \\ v, & \tau = -\tau_{\text{yield}} \text{ and } -s(\tau_{\text{yield}}) \leq v \leq 0 \\ -s(\tau_{\text{yield}}), & \tau = -\tau_{\text{yield}} \text{ and } v < -s(\tau_{\text{yield}}) \\ -s(-\tau), & \tau < -\tau_{\text{yield}}. \end{cases} \qquad (2.5)$$

29

Equations (2.2) – (2.4) may be combined to yield

$$\frac{1}{k}\frac{d\tau}{dt} = v - \frac{dp}{dt} \quad \text{and} \quad m\frac{dv}{dt} = -\tau \tag{2.6}$$

where p satisfies (2.5). Moreover the last identity yields:

$$\frac{1}{2}\frac{d}{dt}[\frac{\tau^2}{k} + mv^2] = -\tau\frac{dp}{dt} \tag{2.7}$$

and

$$\frac{1}{2}[\frac{\tau^2}{k} + mv^2](t) + \int_0^1 \tau\frac{dp}{d\eta}(\eta)d\eta = \frac{1}{2}[\frac{\tau^2}{k} + mv^2](0). \tag{2.8}$$

Equation (2.5) implies that $\tau\frac{dp}{dt}(t) \geq 0$ and that τ and $\frac{dp}{dt}$ are bounded and (2.5) and (2.8) together yield

$$\tau_{\text{yield}}\int_0^\infty |\frac{dp}{d\eta}|(\eta)d\eta \leq \int_0^\infty \tau\frac{dp}{d\eta}d\eta \leq \frac{1}{2}[\frac{\tau^2}{k} + mv^2](0), \tag{2.9}$$

$$\lim_{T\to\infty}\int_T^\infty \phi(\eta)\frac{dp}{d\eta}d\eta = 0 \tag{2.10}$$

for all smooth ϕ's which are bounded in $t \geq T$ and that

$$\lim_{T\to\infty} \text{measure } \{t \geq T \mid |\tau|(t) > \tau_{\text{yield}}\} = 0. \tag{2.11}$$

The inequality (2.9) implies the existence of a permanent plastic deformation p_∞ such that

$$\lim_{t\to\infty} p(t) = p_\infty, \tag{2.12}$$

while (2.10) and (2.11) imply that as t tends to infinity τ and v converge weakly to a solution of the elastic oscillator problem:

$$\frac{1}{k}\frac{d\tilde{\tau}}{dt} = \tilde{v} \quad \text{and} \quad \frac{d\tilde{v}}{dt} = -\tilde{\tau} \tag{2.13}$$

satisfying

$$\frac{1}{2}[\frac{\tilde{\tau}^2}{k} + m\tilde{v}^2](t) \leq \frac{\tau_{\text{yield}}^2}{2k}.^2 \tag{2.14}$$

I shall now turn to the modeling simple shear flows described in the introduction. Here I imagine a continuum undergoing a simple shearing motion; that is material points $\mathbf{X} = X\mathbf{e_1} + Y\mathbf{e_2} + Z\mathbf{e_3}$ move to $\mathbf{x} = X\mathbf{e_1} + (Y + u(X,t))\mathbf{e_2} + Z\mathbf{e_3}$ under the action of the stress tensor $T = \tau(X,t)(\mathbf{e_1} \otimes \mathbf{e_2} + \mathbf{e_2} \otimes \mathbf{e_1}) - \pi(\mathbf{e_1} \oplus \mathbf{e_1} + \mathbf{e_2} \oplus \mathbf{e_2} + \mathbf{e_3} \oplus \mathbf{e_3})$. Defining δ and v by $\delta = \frac{\partial u}{\partial X}$ and $v = \frac{\partial u}{\partial t}$, one finds that compatability of the mixed partial derivatives implies that

$$\frac{\partial \delta}{\partial t} - \frac{\partial v}{\partial X} = 0. \tag{2.15}$$

---

[2]With a little work one can show that if the initial energy satisfies $\frac{1}{2}[\frac{\tau^2}{k} + mv^2](0) > \frac{\tau_{\text{yield}}^2}{2k}$, then after a finite time the solution satisfies $\frac{1}{2}[\frac{\tau^2}{k} + mv^2](t) \equiv \frac{\tau_{\text{yield}}^2}{2k}$.

If one assumes that density in the reference configuration is constant, $\rho_0 > 0$, and that no body forces are present to drive the system. Then balance of momentum reduces to

$$\frac{\partial \pi}{\partial X} = \frac{\partial \pi}{\partial Z} = 0 \quad \text{and} \quad \rho_0 \frac{\partial v}{\partial t} = \frac{\partial \tau}{\partial X} - \frac{\partial \pi}{\partial Y}. \tag{2.16}$$

In the sequel I shall assume that $\frac{\partial \pi}{\partial Y} = 0$, that the shear strain δ may be decomposed into elastic and plastic components e and p respectively; that is

$$\delta = e + p, \tag{2.17}$$

and that the shear stress τ depends linearly on the elastic component e:

$$\tau = ke, \quad k > 0. \tag{2.18}$$

Again, a flow rule is required which describes the evolution of the plastic strain p. The flow rule I choose is motivated by (2.5). I again assume the existence of a monotone nondecreasing function $\tau \to s(\tau)$ on $\tau \geq \tau_{\text{yield}}$ and postulate that

$$\frac{\partial p}{\partial t} = \begin{cases} s(\tau), & \tau > \tau_{\text{yield}} \\ s(\tau_{\text{yield}}), & \tau = \tau_{\text{yield}} \text{ and } \frac{\partial v}{\partial X} > s(\tau_{\text{yield}}) \\ \frac{\partial v}{\partial X}, & \tau = \tau_{\text{yield}} \text{ and } 0 \leq \frac{\partial v}{\partial X} \leq s(\tau_{\text{yield}}) \\ 0, & \tau = \tau_{\text{yield}} \text{ and } \frac{\partial v}{\partial X} < 0 \\ 0, & -\tau_{\text{yield}} < \tau < \tau_{\text{yield}} \\ 0, & \tau = -\tau_{\text{yield}} \text{ and } \frac{\partial v}{\partial X} > 0 \\ \frac{\partial v}{\partial X}, & \tau = -\tau_{\text{yield}} \text{ and } -s(\tau_{\text{yield}}) \leq \frac{\partial v}{\partial X} \leq 0 \\ -s(\tau_{\text{yield}}), & \tau = -\tau_{\text{yield}} \text{ and } \frac{\partial v}{\partial X} < -s(\tau_{\text{yield}}) \\ -s(-\tau), & \tau < -\tau_{\text{yield}}. \end{cases} \tag{2.19}$$

Combing equations (2.15)-(2.18) along with $\frac{\partial \pi}{\partial Y} = 0$ yields

$$\frac{1}{k} \frac{\partial \tau}{\partial t} - \frac{\partial v}{\partial X} = -\frac{\partial p}{\partial t} \quad \text{and} \quad \rho_0 \frac{\partial v}{\partial t} - \frac{\partial \tau}{\partial X} = 0 \tag{2.20}$$

where p satisfies (2.19).

The problem shall report on is the initial value problem for (2.19) and (2.20) on the strip $0 \leq X \leq l$ and $t > 0$. At $t = 0$, I assume that

$$(\tau, v, p)(X, 0^+) = (\tau_0, v_0, p_0)(X), \quad 0 < X < l, \tag{2.21}$$

and at the boundaries that v satisfies

$$v(0, t) = v(l, t) = 0, \quad t > 0.^3 \tag{2.22}$$

My interest is in the long time behavior of weak solutions to (2.19)–(2.22). The existence of such solutions is a relatively trivial matter and is not our concern here.

It is easily checked that such solutions satisfy the energy identity.

$$\frac{1}{2} \int_0^l [\frac{\tau^2}{k} + \rho_0 v^2](X, t) dX + \int_0^t \int_0^l \tau \frac{\partial p}{\partial \eta}(X, \eta) dX d\eta = \frac{1}{2} \int_0^l [\frac{\tau_0^2}{k} + \rho_0 v_0^2](X) dX \tag{2.23}$$

---

[3]Homogeneous boundary conditions for τ at either end are also acceptable.

for all $t \geq 0$. Equation (2.19) implies that for all $t \geq 0$, $\tau(X,t)\frac{\partial p}{\partial t}(X,t) \geq 0$ and that $\tau(\cdot,t)$ and $v(\cdot,t)$ are bounded in $L_2(0,l)$ and (2.19) and (2.23) together yield

$$\tau_{\text{yield}} \int_0^\infty \int_0^l \mid \frac{\partial p}{\partial \eta} \mid (X,\eta)dX d\eta \;\; \leq \;\; \int_0^\infty \int_0^l \tau \frac{\partial p}{\partial \eta}(X,\eta)dX d\eta$$

$$\leq \;\; \frac{1}{2}\int_0^l [\frac{\tau_0^2}{k} + \rho_0 v_0^2](X)dX, \tag{2.24}$$

$$\lim_{T\to\infty} \int_T^\infty \int_0^l \phi \frac{\partial p}{\partial \eta}(X,\eta)dX d\eta = 0 \tag{2.25}$$

for all smooth ϕ's which are bounded in $\{0 \leq X \leq l, T \leq t\}$, and finally that

$$\lim_{T\to\infty} \text{meas} \;\{(X,t) \in [0,l] \times [T,\infty) \mid \;\mid \tau \mid (X,t) > \tau_{\text{yield}}\} = 0. \tag{2.26}$$

Finally, (2.23) and (2.24) imply that if $\tau_0(\cdot)$ and $v_0(\cdot)$ are in $L_2(0,l)$ and $p_0(\cdot)$ is in $L_1(0,l)$, then there exists a permanent plastic deformation $p_\infty(\cdot)$ in $L_1(0,l)$ such that

$$\lim_{T\to\infty} \int_0^l \mid p_\infty(X) - p(X,T) \mid dX = 0. \tag{2.27}$$

In the sequel I shall confine my attention to such data and let $\Omega_+(\tau_0,v_0,p_0)$ be the set of function $(\tau_\infty,v_\infty)(\cdot)$ in $L_2(0,l)\oplus L_2(0,l)$ such that there exits a sequence of times $\{T_n\}_{n=1}^\infty$ with $T_n < T_{n+1}$ and $\lim_{n\to\infty} T_n = \infty$ with the property that the components of the solution to (2.19)–(2.22) at T_n, namely $(\tau,v)(0,T_n)$, converge weakly in $L_2(0,l) \oplus L_2(0,l)$ to $(\tau_\infty,v_\infty)(\cdot)$. (2.27) guarantees that for any such sequence $\{T_n\}_{n=1}^\infty$ the p component of the solution, $p(\cdot,T_n)$, converges to $p_\infty(\cdot)$ strongly in $L_1(0,l)$. The energy estimate (2.23) and the weak compactness of bounded sets in $L_2(0,l)$ guarantee that $\Omega_+(\tau_0,v_0,p_0)$ is nonempty and that the limit functions satisfy

$$\frac{1}{2}\int_0^l [\frac{\tau_\infty^2}{k} + \rho_0 v_\infty^2](X)dX \leq \frac{1}{2}\int_0^l [\frac{\tau_0^2}{k} + \rho_0 v_0^2](X)dX. \tag{2.28}$$

Given such a sequence of times $\{T_n\}_{n=1}^\infty$ and limit function $(\tau_\infty,v_\infty)(\cdot)$ and $p_\infty(\cdot)$ let

$$(\hat{\tau}_n,\hat{v}_n,\hat{p}_n)(\cdot,t) \stackrel{\text{def}}{=} (\tau,v,p)(\cdot,t+T_n), \quad 0 \leq t \leq 2l\sqrt{\frac{\rho_0}{k}} \;^4 \tag{2.29}$$

where $(\tau,v,p)(\cdot,t+T_n)$ is the solution to (2.19)–(2.22) at time $t + T_n$. The basic estimates (2.23) and (2.24) along with the weak compactness of bounded sets in $L_2((0,l)\times(0,2l\sqrt{\frac{\rho_0}{k}})) \oplus L_2((0,l)\times(0,2l\sqrt{\frac{\rho_0}{k}}))$ imply there is no loss in generality in assuming that sequence $(\hat{\tau}_n,\hat{v}_n)(\cdot,\cdot)$ converges weakly in $L_2((0,l)\times(0,2l\sqrt{\frac{\rho_0}{k}})) \oplus L_2((0,l)\times(0,2l\sqrt{\frac{\rho_0}{k}}))$ to limit functions $(\Sigma,V)(\cdot,\cdot)$ and that these functions satisfy (2.28) for almost all $t \in (0,2l\sqrt{\frac{\rho_0}{k}})$. The estimate (2.27) also

---

[4]The number $2l\sqrt{\frac{\rho_0}{k}}$ is the basic temporal period of solutions to the elastic equations

$$\frac{1}{k}\frac{\partial \tau}{\partial t} - \frac{\partial v}{\partial X} = 0 \quad \text{and} \quad \rho_0 \frac{\partial v}{\partial t} - \frac{\partial \tau}{\partial X} = 0$$

satisfying (2.22).

guarantees that the functions $\hat{p}_n(\cdot)$ converge strongly in $L_1((0,l) \times (0, 2l\sqrt{\frac{\rho_0}{k}}))$ to the permanent plastic deformation $p_\infty(\cdot)$. Noting that the functions $(\hat{\tau}_n, \hat{v}_n, \hat{p}_n)(\cdot, \cdot)$ satisfy the identities

$$\iint_B (-\hat{\tau}_n \phi_t + \hat{v}_n \phi_X) dX dt - \int_0^l \phi(X,0)\tau(X, T_n) dX = -\iint_B \phi \frac{\partial \hat{p}_n}{\partial t}(X,t) dX dt \qquad (2.30)$$

and

$$\iint_B (-\hat{v}_n \psi_t + \hat{\tau}_n \psi_X) dX dt - \int_0^l \psi(X,0)v(X, T_n) dX = 0 \qquad (2.31)$$

for all smooth bounded $\phi's$ and $\psi's$ on $B = (0,l) \times (0, 2l\sqrt{\frac{\rho_0}{k}})$ satisfying $\phi(X, 2l\sqrt{\frac{\rho_0}{k}}) = \psi(X, 2l\sqrt{\frac{\rho_0}{k}}) = 0$, $0 \le X \le l$ and $\psi(0,t) = \psi(l,t) = 0$, $0 \le t \le 2l\sqrt{\frac{\rho_0}{k}}$ and noting that (2.25) implies that the right hand side of (2.30) tends to zero as $n \to \infty$ one finds that the limit functions $(\Sigma, V)(\cdot, \cdot)$ are weak solutions of the elastic equations

$$\left. \begin{array}{c} \frac{1}{k}\frac{\partial \Sigma}{\partial t} - \frac{\partial V}{\partial X} = 0 \\ \\ \rho_0 \frac{\partial V}{\partial t} - \frac{\partial \Sigma}{\partial X} = 0 \end{array} \right\} \qquad (2.32)$$

on the strip $0 < X < l$ and $0 < t < 2l\sqrt{\frac{\rho_0}{k}}$ and satisfy

$$(\Sigma, V)(X, 0) = (\tau_\infty, V_\infty)(X), \quad 0 \le X \le l \qquad (2.33)$$

and

$$V(0,t) = V(l,t) = 0, \quad 0 \le t \le 2l\sqrt{\frac{\rho_0}{k}}. \qquad (2.34)$$

The $2l\sqrt{\frac{\rho_0}{k}}$ periodicity of these functions in time guarantees that Σ and V are in fact defined for all times $t \ge 0$. It is also worth noting that these functions satisfy (2.28) for almost all t and that the stress Σ satisfies $-\tau_{\text{yield}} \le \Sigma \le \tau_{\text{yield}}$ almost everywhere. This latter fact follows from (2.26).

The above characterization of the long time behavior of solution to (2.19)–(2.22) is my principal result and concludes the paper.

References

1. Greenberg, J. M., "Models of Elastic-Perfectly Plastic Materials", European Jour. of Appl. Math. (to appear).

2. Antman, S. S. and Szymczak, W. G., "Nonlinear Elastoplastic Wave", Contemporary Mathematics, vol. 100, edited by W. B. Lindquist, AMS Providence, 27-54 (1989).

3. ibid, "Large Antiplane Shearing Motion of Nonlinear Viscoplastic Materials", preprint.

4. Coleman, B.D. and Owen, D.R., "On Thermodynamics and Elastic–Plastic Materials", Arch. Rat. Mech. and Anal. 59, 25–51 (1975).

5. Buhite, J. L., and Owen, D.R., "An Ordinary Differential Equation From the Theory of Plasticity", Arch. Rat Mech. and Anal. 71, 357–383 (1979).

6. Coleman, B.D. and Hodgdon, M.L., "On Shear Bands in Ductile Materials", Arch. Rat. Mech. and Anal. 90, 219–247 (1985).

7. Owen, D.R., "Weakly decaying energy separation and uniqueness of motions of an elastic-plastic oscillator with work–hardening", Arch. Rational Mech. and Anal. 98, 95–114 (1987)

J.M. Greenberg
Department of Mathematics and Statistics
UMBC
Baltimore, MD 21228

A KLARBRING, A MIKELIC AND M SHILLOR
On the rigid punch with friction

We consider the two-dimensional rigid punch with Coulomb's law of dry friction and a normal compliance contact law. The deformable body, represented by a domain $\Omega \subset R^2$, is assumed to be linearly elastic and subject to the body force f and the surface traction h on a part S_h of $\partial\Omega$. It is held fixed on $S_u \subset \partial\Omega$. Moreover, it is assumed that $\partial\Omega = \overline{S}_u \cup \overline{S}_h \cup \overline{S}_c$ and that S_c, S_u and S_h are disjoint and smooth curves. Then for the displacement u and the stress tensor σ we have the following equations

$$\text{div } \sigma + f = 0 \qquad \text{in } \Omega \tag{1}$$

$$\sigma = Ae(u) \quad \text{in } \Omega \tag{2}$$

$$u = 0 \qquad \text{on } S_u, \tag{3}$$

$$\sigma n = h \qquad \text{on } S_h. \tag{4}$$

We suppose $\partial\Omega \in C^{1,1}$, A is a symmetric positive definite fourth order tensor, as usual in linear elasticity and n is the outward unit normal of $\partial\Omega$.

Let us consider the contact boundary S_c. We suppose that

$$S_c = (-a, a) \times \{0\}, \qquad a > 0.$$

On S_c we suppose the following contact law (surface normal compliance):

$$-\sigma_N = c_N[(u_N + \theta x_1 + \beta - g)_+]^m,$$

where $u_N = u \cdot n$, $\sigma_N = \sigma n \cdot n$, $c_N > 0$, $m \geq 1$, β is the vertical downwards displacement of the punch, and θ represents a rotation of the punch around the origin.

In the tangential direction of S_c we assume Coulomb's friction law to hold. More precisely,

$$|\sigma_T| \leq \mu|\sigma_N|, \tag{6}$$

$$|\sigma_T| < \mu|\sigma_N|, \quad \Rightarrow \dot{u}_T - \dot{\alpha} = 0, \tag{7}$$

$$\sigma_T = \mu|\sigma_N|, \quad \Rightarrow \dot{u}_T - \dot{\alpha} \leq 0, \tag{8}$$

$$-\sigma_T = \mu|\sigma_N|, \quad \Rightarrow \dot{u}_T - \dot{\alpha} \geq 0, \tag{9}$$

where $\sigma_T = \sigma n - n\sigma_N$, μ is the coefficient of friction, α is the tangential movement of the punch and $(\cdot)$ denotes the time rate of change.

The derivation of the large displacement contact law and its linearized form are given in [4] and the standard derivation can be found in [1]. Normal compliance contact conditions are described in [6], [2] and [1].

The setting of the problem involves equilibrium equations of the rigid punch. We have

$$N = \int_{-a}^{a} \sigma_N \, dx_1, P = \int_{-a}^{a} \sigma_N \, dx_1 \text{ and } M = - \int_{-a}^{a} x_1 \, \sigma_N \, dx_1,$$

where M is an external moment and N and P are external forces.

From the classical formulation (1)–(9) of the problem of the rigid punch with friction we derive the variational inequality. We define the function space

$$V = \{v \in H^1(\Omega)^2 : v = 0 \text{ on } S_u\} \tag{10}$$

and the bilinear and linear forms

$$a(u,v) = \int_{\Omega} Ae(u) \cdot e(v), \quad \forall\, u, v \in V, \tag{11}$$

$$f(v) = \int_{\Omega} fv + \int_{S_h} hv, \quad \forall\, v \in V. \tag{12}$$

We assume that $f \in L^2(\Omega)^2$, $h \in H^{-\frac{1}{2}}(S_h)$.

Now we introduce the **normal compliance functional**

$$j_N((u,\beta,\theta),(v,\overline{\beta},\overline{\theta})) = \int_{S_c} c_N[(u_N + \theta x_1 + \beta - g)_+]^m \cdot (v_N + \overline{\theta}x_1 + \overline{\beta}),$$

$$\forall(u,\beta,\theta) \in V \times R^2 \text{ and } \forall(v,\overline{\beta},\overline{\theta}) \in V \times R^2, \tag{13}$$

and the **friction functional**

$$j_T((u, \beta, \theta), (v, \overline{\alpha})) = \int_{S_c} \mu c_N [(u_N + \theta x_1 + \beta - g)_+]^m |v_T - \overline{\alpha}|,$$

$$\forall (u, \beta, \theta) \in V \times R^2 \text{ and } \forall (v, \overline{\alpha}) \in V \times R^1. \tag{14}$$

After introducing (11)–(14), we are in a position to formulate our problem.

The Quasistatic Problem

Find $(u, \alpha, \beta, \theta) : [0, T] \rightarrow V \times R^3$ such that $u(0) = u_0 \in V$, $\alpha(0) = \alpha_0$, $\beta(0) = \beta_0$, $\theta(0) = \theta_0$ and $\forall (v, \overline{\alpha}, \overline{\beta}, \overline{\theta}) \in V \times R^3$ there holds

$$a(u, v - \dot{u}) + j_N((u, \beta, \theta), (v, \overline{\beta}, \overline{\theta}) - (\dot{u}, \dot{\beta}, \dot{\theta}))$$

$$+ j_T((u, \beta, \theta), (v, \overline{\alpha})) - j_T((u, \beta, \theta), (\dot{u}, \dot{\alpha})) - M(\overline{\theta} - \dot{\theta})$$

$$- N(\overline{\beta} - \dot{\beta}) - P(\overline{\alpha} - \dot{\alpha}) \geq F(v - \dot{u}). \tag{15}$$

Usually this problem is to be solved by time discretization. After performing an implicit time discretization of this problem, we obtain a sequence of static problems. Following [4] and denoting increments at the level t^ℓ by $u = \Delta u^\ell, \alpha = \Delta \alpha^\ell, \beta = \Delta \beta^\ell, \theta = \Delta \theta^\ell$, we get the variational inequality for $(u, \alpha, \beta, \theta)$. The effects of previous time levels come through the forces which we redefine as

$$F(v) = \int_\Omega fv + \int_{S_h} hv - a(u^{\ell-1}, v), \quad \forall v \in V,$$

and we replace g by

$$g_\ell = g - u^{\ell-1} - \theta^{\ell-1} x_1 - \beta^{\ell-1}.$$

Finally, we get the following variational inequality. Find $(u, \alpha, \beta, \theta) \in V \times R^3$ such that $\forall (v, \overline{\alpha}, \overline{\beta}, \overline{\theta}) \in V \times R^3$ there holds

$$a(u, v - u) + j_N((u, \beta, \theta), (v, \overline{\beta}, \overline{\theta}) - (u, \beta, \theta))$$

$$+ j_T((u, \beta, \theta), (v, \overline{\alpha})) - j_T((u, \beta, \theta), (u, \alpha)) - M(\overline{\theta} - \theta)$$

$$- N(\overline{\beta} - \beta) - P(\overline{\alpha} - \alpha) \geq F(v - u). \tag{16}$$

Therefore, in solving (15) by time discretization we have to solve a problem of the type (16) at every step. It is important to establish existence, uniqueness and regularity results for (16) as a step in solving the quasistatic problem. It is especially important to determine the dual problem for (16) in terms of contact stresses. The derivation of the dual problem (in the sense of Mosco) is discussed in detail in [4]. Here we restrict ourselves only to the presentation of existence and regularity results.

We have the following existence result:

Theorem 1. *Assume* $A \in L^\infty(\Omega)^4$, $c_N \in L^\infty(S_c)$, $c_N \geq c_0 >$, $1 \leq m < \infty$ *and, in addition, that*

$$N > 0, \ |M| < a\,N \ \text{and} \ |P| < \mu N.$$

Then problem (16) has at least one solution.

Proof. The proof is given in [5]. It is established through a fixed point argument, with essential use of the compatibility conditions (17). More precisely, we define an auxilliary problem with a given function G instead of $[(u_N + \theta x_1 + \beta - g)_+]^m$ in the friction functional (14). Then we prove the solvability of the auxilliary problem for every $G \in K$, where

$$K = \{G \in L^{1+1/m}(S_c) \ : \ G \geq 0 \ \text{(a.e.)} \ \text{on} \ S_c, \ \int_{S_c} c_N G = N\}.$$

Moreover, the solution is unique in u, β and θ, and the nonlinear map

$$\Psi(G) = [(u_N + \theta x_1 + \beta - g)_+]^m$$

is well-defined for every $G \in K$. By obtaining appropriate a priori estimates and by using Schauder's fixed point theorem, we prove that the nonlinear map $\Psi : K \to K$ has at least one fixed point. The existence theorems in [2] and [3] for simpler but related problems concerning the elastic body in unilateral contact with a rigid foundation and subject to Coulomb's friction law, are proved by using the general theory of quasi-variational inequalities.

The next proposition gives a necessary condition for existence:

Proposition 2. *Assume that* $A \in L^\infty(\Omega)^4$, $c_N \in L^\infty(S_c)$, $c_N \geq c_0 > 0$ *and* $1 \leq m < \infty$. *If problem (16) is solvable, then*

$$N \geq 0, \qquad |M| \leq a\,N, \qquad |P| \leq \mu N. \tag{18}$$

Proof. It follows from (16) by the use of appropriate test functions.

REMARK 1. The necessary conditions (18) can be interpreted physically as follows. $N \geq 0$ means that the total normal force is pushing the punch against the elastic body. For $N < 0$ the punch is pushed in the opposite direction and equilibrium is impossible. $|M| \leq a\,N$ guarantees that the body does not topple over. $|P| \leq \mu N$ guarantees that the maximal frictional force of resistance can balance the total applied shear force.

REMARK 2. The frictionless punch problem with normal compliance is obtained from (16) by setting $\mu = 0$ and $P = 0$, as equilibrium is impossible with a tangential traction. In this case uniqueness is obvious and existence is a simple consequence of Theorem 1. Our results can be compared with classical results on a frictionless punch in [1] and those with friction in [7].

Now we turn to the regularity of the weak solution for (16):

Theorem 3. *Assume that* $A \in C^2(\overline{\Omega})^4$, $c_N \in C^1(\overline{S}_c)$, $\gamma \in (0,1)$ *and* $g \in C^\gamma(S_c)$ *uniformly in time. Let all assumptions of Theorem 1 be satisfied and let* $(u, \alpha, \beta, \theta) \in V \times R^3$ *be a solution for (16). Then*

$$u \in W^{1+1/p,p}(\Omega_\delta)^2 \cup W^{s,p}(\Omega)^2, \quad \forall p \in [2, \infty), \text{ and } s < \frac{1}{2} + \frac{1}{p},$$

where Ω_δ *is the part of* Ω *remaining after the removal of an arbitrarily small neighbourhood of the points* $(\pm a, 0)$ *and* $\partial S_h \cup \partial S_u$. *Moreover,*

$$u \in C^\lambda(\overline{\Omega})^2, \quad \forall \lambda < \min(\gamma, \frac{1}{2}), \quad \sigma_T(u) \in L^\infty(S_c), \quad \text{and } \sigma_N \in C^\gamma(S_c).$$

Proof. It is given in [5]. It is essentially based on an idea of using the regularity of elements of the subdifferential. We characterize the subdifferential of the friction functional

$$\Phi(v) = \int_{S_c} G|v_T - \overline{\alpha}|, \quad v \in V, \quad \overline{\alpha} \in R$$

and then we reduce the study of the regularity of a solution to (16) to study of the regularity of a mixed boundary value problem in linear elasticity. Of course, the interior regularity is much better.

REFERENCES

1. **N. Kikuchi, J.T. Oden** Contact Problems in Elasticity: A Study of Variational Inequalities and Finite Element Methods, *SIAM,* Philadelphia, 1988.
2. **A. Klarbring, A. Mikelić, M. Shillor** "Frictional contact problem with normal compliance", *Int. J. Engng. Sci.,* 26(8)(1988), 811–832.
3. **A. Klarbring, A. Mikelić, M. Shillor** "On frictional problems with normal compliance", *Nonlin. Anal. TMA,* 13(8) (1989), 935–955.
4. **A. Klarbring, A. Mikelić, M. Shillor** "The rigid punch problem with friction", *Int. J. Engng. Sci.* 29(6) (1991), 751–768.
5. **A. Klarbring, A. Mikelić, M. Shillor** "Mathematical analysis for a problem of the rigid punch with friction", in preparation.
6. **J.T. Oden, J.A.C. Martins** "Models and computational methods for dynamic friction phenomena", *Computation Meth. Appl. Mech. Engng.* 52(1985), 527–634.
7. **D.A. Spence** "An eigenvalue probelm for elastic contact with finite friction", *Proc. Camb. Phil. Soc.* 73(1973), 249–268.

Linköping Institute of Technology, S-58183 Linköping, Sweden

Rudjer Bošković Institute, P.O.B. 1016, 41001 Zagreb, Yugoslavia

Dept. Math. Sciences, Oakland University, Rochester, MI 48309, USA

An elastoplastic problem with frictions

1 Introduction:

We consider a friction problem in the geometrically linear elastoplasticity (Hencky's plasticity) theory. We establish an appropriate variational formulation and prove the existence result by "relaxing" the problem reasonably.

The idea of relaxing comes from the well known fact that the strain solutions of the Hencky's plasticity problems are not in the Sobolev spaces. A cosequence of this is that the displacement boundary condition is NOT preserved. In the friction problems, the same thing happens: due to plasticity, we can not "hold" the material tightly on the boundary to satisfy the displacement boundary condition imposed, so neither should we expect that we can "hold" the material on the boundary to satisfy Comlomb's law. Therefore, the existence result should also be established on some kind of relaxation principal.

2 Mathematical Formulation of the Problem:

To establish the mathematical formulation of Coulomb's law in linearized elasticity-plasticity problems, we follow formally the friction theory and the elasticity-plasticity theory to give the following formulation: Find $u \in H^1(\Omega; R^3)$, $\sigma \in L^2(\Omega; E)$ such that

$$e(u) = A\sigma + \lambda \text{ in } \Omega, \tag{1}$$

$$div\sigma + f(x) = 0 \text{ in } \Omega, \tag{2}$$

$$\lambda : (\phi - \sigma) \leq 0 \text{ a.e. in } \Omega \text{ for all } \phi \in L^2(\Omega; E) \,,$$

$$\text{such that } |\phi^D(x)| \leq k \text{ a.e.,} \tag{3}$$

$$u|_{\Gamma_1} = u_0, \tag{4}$$

$$\sigma n|_{\Gamma_2} = g, \tag{5}$$

$$(\sigma n)_n|_{\Gamma_3} = F_n, \tag{6}$$

$$|(\sigma n)_\tau| < l \Rightarrow u_\tau = 0 \text{ on } \Gamma_3 \,, \tag{7}$$

$$|(\sigma n)_\tau| = l \Rightarrow \exists q \geq 0 \text{ such that } u_\tau = -q(\sigma n)_\tau \text{ on } \Gamma_3 \,. \tag{8}$$

where Ω is a bounded, open, C^2 subset of R^3; E is $R^{3 \times 3}_{sym}$ and $E^D = \{\xi \in E : \xi_{11} + \xi_{22} + \xi_{33} = 0\}$; e is the linearized deformation operator from $H^1(\Omega; R^3)$ into $L^2(\Omega; E)$ such that for any $u \in H^1(\Omega; R^3)$,

$$e_{ij}(u) = (\frac{\partial u_i}{\partial x_j} + \frac{\partial u_j}{\partial x_i})/2 \text{for } i, j = 1, 2, 3;$$

A is a linear operator from E into E such that for any $\sigma \in E$,

$$A\sigma = b_1 \sigma^D + b_2 tr\sigma I$$

with b_1, $b_2 > 0$, $tr\sigma = \sigma_{11} + \sigma_{22} + \sigma_{33}$, $I = $ identity in E and $\sigma^D = \sigma - \frac{1}{3}tr\sigma I$; Γ_1, Γ_2, Γ_3 are C^1 open subsets of Γ, $\bar{\Gamma}_1 \cup \bar{\Gamma}_2 \cup \bar{\Gamma}_3 = \Gamma$, $\Gamma_i \cap \Gamma_j$ is empty for all $i \neq j$; l and k are positive physical constants; u_0 is the trace of a H^1 function; $g \in L^\infty(\Gamma_2; R^3)$; F_n in $W^{1/p, p/(p-1)}(\Gamma_3)$ for some $p \in (1, 3/2)$; $f \in L^\infty(\Omega; R^3)$.

To simplify the arguments in the following, we suppose that Ω is connected. We will write simplify $W^{s,p}(\Omega; R^3)$ as $W^{s,p}(\Omega)$ in the following when there is no confusion.

3 Variational formulation and Limit Analysis problems:

We transform (1)-(8) into a pair of formally equivalent variational problems:

$$\mathcal{P} : Inf\{\int_\Omega \psi(e(u))dx + \int_{\Gamma_3} l|u_\tau|d\Gamma - \int_\Omega f(x)u(x)dx -$$

$$- \int_{\Gamma_2} u \cdot g d\Gamma - \int_{\Gamma_3} F_n(u \cdot n)d\Gamma\} \tag{9}$$

subject to $u \in A = \{u \in H^1(\Omega), u|_{\Gamma_1} = u_0\}$ and

$$\mathcal{P}^* : Sup\{-\int_\Omega A\sigma : \sigma dx + \int_{\Gamma_1} (\sigma \cdot n)u_0 d\Gamma\} \tag{10}$$

subject to $\sigma \in A^* = \{\sigma \in L^2(\Omega; E), |\sigma^D(x)| \leq k \text{ a.e.}, div\sigma(x) + f(x) = 0, \sigma n|_{\Gamma_2} = g,$
$(\sigma n)_n|_{\Gamma_3} = F_n, |(\sigma n)_\tau|_{\Gamma_3}| \leq l.\}$ with

$$\psi(\xi) = Sup\{\xi : \eta - A\eta : \eta; \quad \eta \in E, |\eta^D| \leq k.\} \tag{11}$$

for all $\xi \in E$ and we have the following

Proposition 3.1: If u is a solution of $\mathcal{P}$, σ a solution of $\mathcal{P}^*$, then (u, σ) satisfies the relations (1)-(8) and vice versa.

Propositon 3.2: The function ψ defined in (11) satisfies the following properties:

$$c_1(tr\xi)^2 + c_3(|\xi^D| - 1) \leq \psi(\xi) \leq c_2(tr\xi)^2 + c_4(|\xi^D| + 1), \tag{12}$$

$$\psi(\xi) \geq 0, \psi(0) = 0, \tag{13}$$

with c_1, c_2, c_3, c_4 being positive constants (cf. [17]).

Proposition 3.3: $Inf\mathcal{P} = Sup\mathcal{P}^*$ (cf. [9]).

It follows from Proposition 3.2 that ψ has only linear growth at infinity with respect to $\xi \in E^D$. So in general, we do not have any result ensuring that $Inf\mathcal{P} > -\infty$. Therefore, a condition which can guarantee that the functional is bounded below is desirable. One such condition called the limit analysis hypothesis was given via the following problem:

$$PLA : Inf\{\int_\Omega \psi_\infty(e^D(u))dx + \int_{\Gamma_3} l|u_\tau|d\Gamma\} \tag{14}$$

subject to $u \in AL = \{u \in H^1(\Omega), u|_{\Gamma_1} = 0, \int_\Omega f(x)u(x) + \int_{\Gamma_2} g \cdot ud\Gamma + \int_{\Gamma_3} F_n(u \cdot n)d\Gamma = 1, divu = 0.\}$ with $\psi_\infty(\xi^D) = lim_{t \to \infty}\psi(t\xi^D)/t$ for any $\xi \in E$.

4 Function Spaces and Limit Analysis Hypothesis:

Definitions 4.1: Let $M(\Omega)$ denote the space of bounded Radon measures and we define

$$BD(\Omega) = \{u \in L^1(\Omega), e(u) \in M(\Omega; E)\},$$

$$U(\Omega) = \{u \in BD(\Omega), divu \in L^2(\Omega)\}$$

with the following norms

$$|\mu|_{M(\Omega)} = Sup\{\int_\Omega \phi\mu; \phi \in C_0(\Omega), |\phi(x)| \leq 1\},$$

43

$$|u|_{BD(\Omega)} = |u|_{L^1} + |e(u)|_{M(\Omega;E)},$$
$$|u|_{U(\Omega)} = |u|_{BD(\Omega)} + |divu|_{L^2(\Omega)}.$$

Definitions 4.2: The "weak" topology of

i) $BD(\Omega)$ *is defined as:* $u_n \rightharpoonup u$ *in* $BD(\Omega)$ *if and only if* $u_n \rightarrow u$ *in* $L^1(\Omega)$ *and* $e(u_n) \rightarrow e(u)$ *in* $M(\Omega;E)$ *weak-star.*

ii) $U(\Omega)$ *is defined as:* $u_n \rightharpoonup u$ *in* $U(\Omega)$ *if and only if* $u_n \rightharpoonup u$ *in* $BD(\Omega)$ *and* $divu_n \rightharpoonup divu$ *in* $L^2(\Omega)$ *(in usual* L^p *spaces, "$\rightharpoonup$" denotes weak convergence).*

From the above definitions, we can draw the following consequences: $U(\Omega)$ embeded continuously in $BD(\Omega)$; the embeding $BD(\Omega) \hookrightarrow L^p(\Omega)$ is continuous for any $p \in [1, 3/2]$ and is compact for any $p \in [1, 3/2)$; from any bounded sequence of $BD(\Omega)$ or $U(\Omega)$, we can extract a subsequence which converges weakly in $BD(\Omega)$ or $U(\Omega)$ respectivly.

Now, as all the proper function spaces have been defined, we give explicitly the limit analysis hypothesis and make clear its mathematical significance.

Hypothesis 4.4: $Inf\mathcal{P}LA > 1.$

Proposition 4.5: Under Hypo 4.4, the problem $\mathcal{P}^*$ *admits a unique solution and any minimizing sequence of* $\mathcal{P}$ *will stay in a bounded set of* $U(\Omega)$.

5 Relaxation:

In the preceding section, we showed that under Hypo 4.4, the minimizing sequence of $\mathcal{P}$ is bounded in $U(\Omega)$ and therefore, containing a subsequence which converges in $U(\Omega)$ weakly. We want to show that the limit function obtained is a solution of the problem. To show this, we have two things to explain and verify. The first is that the weak convergence in $U(\Omega)$ does not guarantee that the boundary condition can be maintained. So we have to find a compromise to say that the limit function is a solution in some sense. The second is that following that compromise, we have to justify that the functional considered is l.s.c. with respect to the convergence available.

We recall the main facts know for the first case here in adapting to our situation. The boundary condition is relaxed to $u \cdot n|_{\Gamma_1} = u_0 \cdot n$ and the problem is relaxed to

$$\mathcal{PR} : Inf\{\int_{\Omega} \psi(e(u)) + \int_{\Gamma_1} \psi_{\infty}(\mathcal{T}^D(u_{0\tau} - u_{\tau}))d\Gamma + \int_{\Gamma_3} l|u_{\tau}|d\Gamma$$

$$- \int_\Omega f(x)u(x)dx - \int_{\Gamma_2} u \cdot g d\Gamma - \int_{\Gamma_3} F_n(u \cdot n)d\Gamma \qquad (15)$$

subject to $u \in AR = \{u \in H^1(\Omega), \ u \cdot n|_{\Gamma_1} = u_0 \cdot n\}$, where T is the linear operator from $L^1(\partial\Omega)$ into $L^1(\partial\Omega; E)$ defined by $T(u)_{ij} = (u_i n_j + u_j n_i)/2$. We know that $\mathcal{PR}$ is a relaxation of $\mathcal{P}$ in the following sense:

i) $\mathrm{Inf}\mathcal{PR} = \mathrm{Inf}\mathcal{P}$,

ii) any minimizing sequence of $\mathcal{P}$ is also a minimizing sequence of $\mathcal{PR}$.

Let us define

$$C = \{\xi \in E : \ \forall n \in R^3, |n| = 1, |\xi n - (n^T \xi n)n| \le l\}. \qquad (16)$$

Proposition 5.1: $C = RI + C^D$ with C^D a bounded convex subset of E^D containing a neighbourhood of 0.

In the following, we denote $\psi_1(\xi) = Sup\{\xi : \eta - A\eta : \eta; \ \eta \in C\}$ and $\tilde{\psi}(\xi) = the convexification of $min\{\psi_1(\xi), \psi(\xi)\}$ and we have

Proposition 5.2: $\tilde{\psi}(\xi)$ satisfies the estimates (12) and (13).

Proposition 5.3: For any $u \in H^1(\Omega)$, we have

$$\int_{\Gamma_3} \psi_{1\infty}(-T^D(u_\tau))d\Gamma = \int_{\Gamma_3} l|u_\tau|d\Gamma. \qquad (17)$$

We want $\tilde{\psi}(-T^D(u_\tau))$ to be the relaxation of $l|u_\tau|$ in the expression of potential of energy and so we try to give its explicit expression.

Proposition 5.4: For any $u \in H^1(\Omega)$, we have

$$\int_{\Gamma_3} \tilde{\psi}_\infty(-T^D(u_\tau))d\Gamma = Sup \int_{\Gamma_3} -\sigma : T^D(u_\tau)d\Gamma \qquad (18)$$

subject to $\sigma \in \hat{C}$ (see (24)).

We can now give the relaxed problem of $\mathcal{PR}$ and show that it does not influence the minimum value of energy.

$$\mathcal{QR} : Inf\{\int_\Omega \psi(e(u)) + \int_{\Gamma_1} \psi_\infty(T^D(u_{0\tau} - u_\tau))d\Gamma + \int_{\Gamma_3} \tilde{\psi}_\infty(-T^D(u_\tau))d\Gamma$$

$$- \int_\Omega f(x)u(x)dx - \int_{\Gamma_2} g \cdot u d\Gamma - \int_{\Gamma_3} F_n(u \cdot n)d\Gamma\} \qquad (19)$$

subject to $u \in AR$. *When we compare* $\mathcal{QR}$ *with* $\mathcal{PR}$, *it is easy to see that any addmissible function of* $\mathcal{PR}$ *is one of* $\mathcal{QR}$ *and* $Inf\mathcal{PR} \geq Inf\mathcal{QR}$. *So we have to show that*

$$Inf\mathcal{PR} = Inf\mathcal{QR}. \tag{20}$$

By the generalized duality (cf [11]), for any u *which is* $\mathcal{PR}$ *admissible,* σ *which is* $\mathcal{P}^*$ *admissible, we have*

$$\int_\Omega \psi(e(u)) + \int_{\Gamma_1} \psi_\infty(\mathcal{T}^D(u_{0\tau} - u_\tau))d\Gamma + \int_{\Gamma_3} \tilde{\psi}_\infty(-\mathcal{T}^D(u_\tau))d\Gamma - \int_\Omega f(x)u(x)dx$$

$$- \int_{\Gamma_2} g \cdot u d\Gamma - \int_{\Gamma_3} F_n(u \cdot n)d\Gamma \geq - \int_\Omega A\sigma : \sigma dx + \int_{\Gamma_1} (\sigma n) \cdot u_0 d\Gamma$$

which implies that $Inf\mathcal{QR} \geq Sup\mathcal{P}^* = Inf\mathcal{PR}$. *So (29) holds. We now extend the definition domain of the problem.*

$$\mathcal{Q} : Inf\{\int_\Omega \psi(e(u)) + \int_{\Gamma_1} \psi_\infty(\mathcal{T}^D(u_{0\tau})) + \int_{\Gamma_3} \tilde{\psi}_\infty(-\mathcal{T}^D(u_\tau))$$

$$- \int_\Omega f(x)u(x)dx - \int_{\Gamma_2} g \cdot u d\Gamma - \int_{\Gamma_3} F_n(u \cdot n)d\Gamma\}$$

subject to $u \in \{u \in U(\Omega),\ u \cdot n|_{\Gamma_1} = u_0 \cdot n\}$. *The relation between* $\mathcal{Q}$ *and* $\mathcal{P}$ *is that* $Inf\mathcal{P} = Inf\mathcal{Q}$ *and any admissible function of* $\mathcal{P}$ *is one of* $\mathcal{Q}$.

6 Existence Theorem in Strain Problem:

In this section, we show that for any sequence $\{u_m\} \subset U(\Omega)$ *which converges weakly in* $U(\Omega)$ *to* u, *with each* u_m *being* $\mathcal{Q}$ *admissible, we have*

$$\underline{lim}_{m\to\infty}\{\int_\Omega \psi(e(u_m)) + \int_{\Gamma_1} \psi_\infty(\mathcal{T}^D(u_{0\tau} - u_{m\tau}))d\Gamma + \int_{\Gamma_3} \tilde{\psi}_\infty(-\mathcal{T}^D(u_{m\tau}))d\Gamma$$

$$- \int_\Gamma f(x)u_m(x)dx - \int_{\Gamma_2} g \cdot u_m d\Gamma - \int_{\Gamma_3} F_n(u_m \cdot n)d\Gamma\}$$

$$\geq \int_\Omega \psi(e(u)) + \int_{\Gamma_1} \psi_\infty(\mathcal{T}^D(u_{0\tau} - u_\tau))d\Gamma + \int_{\Gamma_3} \tilde{\psi}_\infty(-\mathcal{T}^D(u_\tau))d\Gamma$$

$$- \int_\Omega f(x)u(x)dx - \int_{\Gamma_2} g \cdot u d\Gamma - \int_{\Gamma_3} F_n(u \cdot n)d\Gamma. \tag{21}$$

Theorem 6.1: If (30) is true, then under Hypo 4.4, the problem $\mathcal{Q}$ *admits at least one solution.*

Lemma 6.2: If ψ_i: $E^D \to R^+ \bigcup\{0\}$ *for* $i = 1, 2$ *such that*
i) $c_i(|p| - 1) \leq \psi_i(p) \leq c_i'(|p| + 1)$ *with* $c_i, c_i' > 0$, $p \in E^D$, $i = 1, 2$.
ii) ψ_i *is convex.*

iii) $\psi_2(p) \leq \psi_1(p)$.

Then, if Ω is a bounded regular open subset of R^3 with boundary Γ, Γ_0 is a regular closed subset of Γ and if $\{\mu_n\} \subset M(R^3; E^D)$ such that

i) supp $\mu_n \subseteq \bar{\Omega}$,

ii) $\mu_n \to \mu$ in $M(R^3; E^D)$ weak-star,

then we have

$$\underline{lim}_{n\to\infty} \int_{\Omega} \psi_1(\mu_n) + \int_{\Gamma_0} \psi_{2\infty}(\mu_n|_{\Gamma_0}) \geq \int_{\Omega} \psi(\mu) + \int_{\Gamma_0} \psi_{2\infty}(\mu|_{\Gamma_0}),$$

with $\mu|_{\Gamma_0}$ the part of measure of μ supported by Γ_0 for any $\mu \in M(R^3, E^D)$.

Theorem 6.3: (30) holds under the hypotheses given at the begining of the section.

Tang Qi
Department of Mathematics
Heriot-Watt University
Ricarton
Edinburgh EH14 4AS
U.K.

B D REDDY

Aspects of the modern theory of elastoplasticity

Abstract

The classical theory of elastoplasticity has undergone extensive change during the last twenty years. The mathematical basis of the theory has benefited from efforts to place it within the framework of convex analysis, in that a clearer, more unified theory has emerged. The qualitative theory of the partial differential equations and inequalities of elastoplasticity has also seen considerable progress so that, though this theory cannot be said to be complete, what remains to be done is not beyond the reach of analysts.

We first review recent work [1] which has for the first time provided a comprehensive convex analytic framework for small strain elastoplasticity. The second part of the contribution focuses on recent work on the existence theory for quasi-static problems in plasticity. It is shown firstly how a formulation other, more classical formulations of the problem, is described. We present an existence theory for the problem thus formulated. This theory is 'semi-constructive', in that it suggests algorithms for the solution of the problem by computational means.

1   Introduction

The aim of this contribution is to report on some recent developments in the modern theory of elastoplasticity. First, we review recent work [1] which provides a comprehensive convex analytic framework for small strain elastoplasticity; in particular, this work establishes the precise nature of the maximum plastic work inequality, as well as the manner in which it is equivalent to alternative formulations of the evolution or flow law. The second part of the contribution focuses on recent work on the existence theory for quasi-static problems in plasticity. It is shown firstly how a formulation in which the support function (or dissipation function) is used, is more natural, in the sense that the problem is a natural extension of the standard displacement problem of elasticity. We then discuss an existence theory for the variational problem thus formulated.

2 Elastoplastic constitutive relations

We focus attention on small-strain isothermal elastoplasticity and for convenience consider quasi-static deformations, in which inertial terms may be neglected. Under these circumstances the form which the constitutive equations take is governed by the dissipation inequality

$$\rho\dot{\psi} - \sigma.\dot{\varepsilon} \leq 0 \tag{2.1}$$

where σ is the symmetric stress tensor, ε the strain tensor, ρ the mass density, b the body force, and ψ the free energy. We consider a class of constitutive equations of the form

$$\psi = \bar{\psi}(\varepsilon, \lambda), \tag{2.2}$$

$$\sigma = \bar{\sigma}(\varepsilon, \lambda), \tag{2.3}$$

$$\dot{\lambda} = F(\varepsilon, \lambda), \tag{2.4}$$

in which λ is an ensemble of internal variables. Substitution of (2.2) and (2.3) in the dissipation inequality (2.1) yields, in the usual way,

$$\sigma = \rho\frac{\partial\bar{\psi}}{\partial\varepsilon} \quad \text{and} \quad \chi.\dot{\lambda} \geq 0 \tag{2.5}$$

where χ is the force conjugate to λ, defined by

$$\chi = \bar{\chi}(\varepsilon, \lambda) := -\frac{\partial\bar{\psi}}{\partial\lambda}. \tag{2.6}$$

For conventional rate-independent models of plasticity it is not unduly restrictive to suppose that in the evolution law (2.4), $\dot{\lambda}$ depends on ε and λ through χ, an assumption which we adopt so that we write

$$\dot{\lambda} = F(\chi). \tag{2.7}$$

In determining the form which F should take we are guided by the notion that the theory should include as a special case the classical theory of plasticity based on the assumption of the existence of a closed convex set of admissible stresses, together with the assumption that the plastic strain rate lies in the normal cone at any point of this set. The inequality $(2.5)_2$ also is a reminder of the dual roles which χ and $\dot{\lambda}$ should play in a convex analytic framework. With all of this in mind we postulate the existence of a closed convex set $K \subset E = R^{n \times n}$ with the property that $0 \in K$, and we assume that (2.7) takes the form

$$\chi \in K, \qquad \dot{\lambda} \in N(\chi) \Leftrightarrow \chi \in \partial D(\dot{\lambda}), \tag{2.8}$$

where $N(\chi)$ is the normal cone to K at χ and D is the support function of K, defined by

$$D : E^* \to R \cup \{+\infty\}, \quad D(p) = \sup_{\tau \in K} \tau . p. \tag{2.9}$$

The relation between $(2.5)_2$ and the support function is clear if we bear in mind that D achieves its supremum at χ. In (2.8), ∂D denotes the subdifferential of D and (2.8) is simply an expression of the fact that D and I, the indicator function of K, are Legendre-Fenchel conjugates [2]. This form of the elastoplastic law is not new (see, for example, Halphen and Nguyen [3] for an earlier treatment), but there is a further equivalent form which has recently been discussed [1], in an attempt to clarify the relationship between (2.8) and the classical maximum plastic work inequality [4]. For this purpose we define a multivalued map $G : p \to G(p)$ to be *responsive* if $0 \in G(0)$ and if for any p_0, p_1, $\in E$,

$$(\chi_0 - \chi_1).p_0 \geq 0 \quad \text{and} \quad (\chi_1 - \chi_0).p_1 \geq 0$$

whenever $\chi_0 \in G(p_0)$ and $\chi_1 \in G(p_1)$. The map G is said to be *maximal responsive* if it is responsive and if there is no other responsive map whose graph properly includes the graph of G. The main result of Eve, Reddy and Rockafellar [1], when specialised to our particular application answers the question whether, given a multivalued map of G, there exists a function D (eg the support function of K in our case) such that

$$G = \partial D \tag{2.10}$$

(the converse is of course trivial). We have

THEOREM 1. *G is a maximal responsive map if and only if there exists a lower semicontinuous gauge D with the property (2.10).* □

Thus the evolution law can be expressed in the three equivalent forms

$$\dot{\lambda} \in N(\chi) \Leftrightarrow \chi \in \partial D(\dot{\lambda}) \Leftrightarrow \chi \in G(\dot{\lambda}).$$

3 The quasistatic initial-boundary value problem for elastoplastic media

For convenience we restrict attention to bodies having a quadratic free energy function given by

$$f(\varepsilon, \lambda) = \tfrac{1}{2}\varepsilon.C\varepsilon + \varepsilon.E\lambda + \tfrac{1}{2}\lambda.H\lambda$$

in which C, E and H are tensors of material properties. Then from (2.5) and (2.6) we have

$$\bar{\sigma}(\varepsilon, \lambda) \;=\; C\varepsilon + E\lambda, \tag{3.1}$$

$$-\bar{\chi}(\varepsilon, \lambda) \;=\; E^T\varepsilon + H\lambda. \tag{3.2}$$

We adopt the evolution law in the form

$$\chi \in \partial D(\dot{\lambda}), \tag{3.3}$$

where D is a given lower semicontinuous gauge, and consider

PROBLEM 1. Suppose that the body occupies a region Ω in R^3 with boundary Γ and suppose that the history of body force $b(x,t)$ is given on Ω. For simplicity we assume that the displacement field, denoted by $u(x,t)$, is zero on Γ. We are required to find $u(x,t)$ and $\lambda(x,t)$ which satisfy the equation of equilibrium

$$div \ \sigma + \rho b = 0 \ \ in \ \Omega, \tag{3.4}$$

the strain-displacement relation

$$\varepsilon(u) = \tfrac{1}{2}\left(\nabla u + (\nabla u)^T\right), \tag{3.5}$$

the constitutive relations (3.1) - (3.3) in Ω, and the initial conditions

$$u(x,0) = \lambda(x,0) = 0. \tag{3.6}$$

It is worth noting that the formulation just given is rather different from that commonly considered in mathematical studies of the elastoplastic problem. We have in mind here the works of Duvaut and Lions [5] and of Johnson [6] as prime examples of an alternative approach. The formulation adopted here has the following advantages: it specialises in a straightforward way to the standard displacement problem of linear elasticity; unlike the formulations of Duvaut and Lions and of Johnson in which *stress* and *velocity* are sought, the main variables here are *displacement* (a far more natural variable) and the internal variables; finally, the present formulation is a useful one for studying in parallel problems of perfect plasticity and of hardening plasticity, two classes of problems which require distinct treatments from a mathematical point of view [8].

At this stage it is necessary to be specific about the nature of the internal variable λ if we are to make any progress in the analysis of the problem. For simplicity we choose λ to be the *plastic strain*; the elastic relation (3.1) immediately dictates then that $L = -C$. We leave H unspecified for the present.

The spaces of functions in which solutions are sought are denoted V and Λ, where

$$V = H_0^1(\Omega)^n \quad \text{and} \quad \Lambda = \left\{ \mu : \mu_{ji} = \mu_{ij}, \quad \mu_{ii} = 0, \quad \mu_{ij} \in L^2(\Omega) \right\}. \tag{3.7}$$

The definition $(3.7)_2$ encompasses the requirements that the plastic strain be symmetic and that the volume change associated with plastic deformation be zero. Λ is clearly a closed subspace of $L^2(\Omega)^{n \times n}$. Note that, here and henceforth, summation is implied on repeated indices.

We are now ready to formulate the problem in weak form. The equation of equilibrium (3.4), the strain-displacement relation (3.5) and the elastic relation (3.1) give, after the usual manipulation, the equation

$$\underbrace{\int_\Omega C\varepsilon(u).\varepsilon(v)dx}_{a(u,v)} - \underbrace{\int_\Omega C\lambda.\varepsilon(v)dx}_{b(v,\lambda)} = \underbrace{\int_\Omega b.v \, dx}_{<l,v>}, \qquad v \in V,$$

the bilinear forms $a : V \times V \to R$ and $b : V \times \Lambda \to R$, and linear functional $l : V \to R$ being defined in an obvious way. The evolution law (3.3) together with the expression (3.2) for χ give, after using the definition of the subdifferential, the inequality

$$j(\mu) - j(\dot{\lambda}) - b(u, \mu - \dot{\lambda}) + c(\lambda, \mu - \dot{\lambda}) \geq 0, \qquad \mu \in \Lambda,$$

in which the functional $j : \Lambda \to R$ and bilinear form $c : \Lambda \times \Lambda \to R$ are defined by

$$j(\mu) = \int_\Omega D(\mu)dx, \quad c(\lambda, \mu) = \int_\Omega H\lambda.\mu \, dx.$$

We now have

PROBLEM 2. Find $(u, \lambda) : [0, T] \to V \times \Lambda$ such that $a.e.$ on $[0, T]$,

$$a(u, v) - b(v, \lambda) = <l, v>, \quad v \in V, \tag{3.8}$$

53

$$c(\lambda, \mu - \dot{\lambda}) + j(\mu) - j(\dot{\lambda}) - b(u, \mu - \dot{\lambda}) \; \geq \; 0, \quad \mu \in \Lambda, \tag{3.9}$$

$$u(0) = \lambda(0) \; = \; 0.$$

Our main result is then embodied in

THEOREM 2. *Suppose that $b_i \in W^1(0, T; L^2(\Omega))$, $C_{ijkl} \in L^\infty(\Omega)$, $H_{ijkl} \in L^\infty(\Omega)$. Suppose further that C and H exhibit the symmetries $ij \leftrightarrow kl$, $ji \leftrightarrow ij$ and $lk \leftrightarrow kl$, and that there exist constants $\alpha > 0$ and $\beta > 0$ such that*

$$C_{ijkl}\, M_{ij}\, M_{kl} \geq \alpha M_{ij}\, M_{ij}, \quad (H_{ijkl} - C_{ijkl})\, M_{ij}\, M_{kl} \geq \beta M_{ij} M_{ij} \tag{3.10}$$

for all symmetric matrices M. Then Problem 2 has a unique solution $(u, \lambda) \in L^\infty(0, T; V \times \Lambda)$, with $(\dot{u}, \dot{\lambda}) \in L^2(0, T; V \times \Lambda)$. □

Remarks.

1. The proof of Theorem 2 is lengthy, and is presented elsewhere [7]. It is approached by constructing an approximate solution (u_k, λ_k) based on a finite-difference discretisation in time of $(u(t), \lambda(t))$, and then by establishing certain *a priori* estimates independent of the time-step k. The final step involves passage to the limit.

2. It is interesting to consider some typical situations in which Theorem 2 can be applied. For example, in the case of von Mises plasticity with linear kinematic hardening we have

$$D(\mu) = k\sqrt{\mu_{ij}\mu_{ij}} \quad \text{and} \quad H = C + hI, \tag{3.11}$$

in which k and h are bounded and measurable scalars which are strictly positive almost everywhere. The conditions of the Theorem clearly hold provided that the elastic moduli C satisfy the ellipticity condition $(3.10)_1$. On the other hand, in the case of *perfect* plasticity we have $h = 0$ and are unable to deduce anything from the Theorem. This is as it should be, since it is now well-known that solutions to perfectly plastic problems lie not in Sobolev spaces, but in the space of functions of bounded deformation $BD(\Omega)$ (for displacements) and of bounded measures $\mathcal{M}$ (for plastic strains) [8].

Acknowledgement

This work has been supported by the Foundation for Research Development.

References

[1] R A Eve, B D Reddy and R T Rockafellar, An internal variable theory of plasticity based on the maximum plastic work inequality. *Quart Appl Math* **48** (1990) 59-84.

[2] R T Rockafellar, *Convex Analysis*. Princeton University Press (Princeton) (1970).

[3] B Halphen and Q S Nguyen, Sur les materiaux standards generalisés. *Jour de Méc* **14** (1975) 39-63.

[4] J B Martin, *Plasticity: Fundamentals and General Results*. MIT Press (Cambridge, Mass.) (1975).

[5] G Duvaut and J L Lions, *Inequalities in Mechanics and Physics*. Springer (Berlin) (1972).

[6] C Johnson, Existence theorems for plasticity problems. *Jour Math Pure et Appl* **55** (1976) 431-444.

[7] B D Reddy, Existence of solutions to a quasistatic problem in elastoplasticity. To appear in *Elliptic and Parabolic Problems*. (ed J Bemelmans and M Grüter) Longman (London) (1992).

[8] B D Reddy and R Tomarelli, The obstacle problem for an elastic-plastic body. *Appl Math Opt* **29** (1990) 89-110.

B D Reddy
Department of Applied Mathematics, University of Cape Town, 7700 Rondebosch, South Africa

T ROUBICEK

A Stefan problem in multi-component media coupled with a Signorini problem

This contribution deals with the nonlinear heat transfer problem, admitting also the Stefan problem, in a medium composed of, e.g., two homogeneous subdomains Ω_1, Ω_2 in a unilateral contact on the boundary $\Gamma_c = \bar{\Omega}_1 \cap \bar{\Omega}_2$. Moreover, we will consider a standard thermo-visco-elastic problem (in terms of small deformations) in each subdomain. Beside a usual temperature dilatation, we will couple the temperature field with the displacement one backward via a displacement dependence of the heat resistivity coefficient that appears in the contact conditions for the heat transfer problem – typically, the larger the slot between the subdomains, the greater the heat resistivity coefficient. Such a problem appears in various applications: in foundry industry within solidification of a molten ingot in a form, in geophysics within modelling of plate tectonics, etc. It also generalizes, in some directions, the problem from [4].

1. Enthalpy formulation of the problem

For functions $\alpha_l, \beta_l : \mathbb{R} \to \mathbb{R}$ describing heat transfer properties of the material ocupying l^{th} subdomain, $l = 1, 2$ (cf. [5]), we consider the heat transfer equations:

$$(1) \qquad \begin{cases} \theta(x,t) = \alpha_l(w(x,t)) \\ \partial w / \partial t = \Delta \beta_l(w) \end{cases} \quad \text{on } Q_l = \Omega_l \times (0,T), \quad l = 1, 2,$$

with the contact conditions

$$(2) \qquad \frac{\partial}{\partial \nu} \beta_1(w^{(1)}) = \frac{\partial}{\partial \nu} \beta_2(w^{(2)}) \quad \text{on } \Sigma_c = \Gamma_c \times (0,T),$$

$$(3) \qquad \alpha_1(w^{(1)}) - \alpha_2(w^{(2)}) = r([u]\nu) \frac{\partial}{\partial \nu} \beta_1(w^{(1)}) \quad \text{on } \Sigma_c.$$

the boundary conditions on the remaining parts of the boundaries of Ω_l, i.e. on $\Gamma_l = \partial \Omega_l \setminus \Gamma_c$,

$$(4) \qquad \frac{\partial}{\partial \nu} \beta_l(w) + g_l(x,t,\alpha_l(w)) = 0 \quad \text{on } \Sigma_l = \Gamma_l \times (0,T), \quad l = 1, 2,$$

and finally the initial condition

$$(5) \qquad w(.,0) = w^0(.) \quad \text{na } \Omega = \Omega_1 \cup \Omega_2.$$

In the above equations, w denotes the enthalpy, θ temperature, ν the outward unit normal to Ω_l in the case of (4) or the unit normal oriented from Ω_2 to Ω_1 in the other cases, g_l a prescibed heat flux, w^0 an initial enthalpy, and r the heat resistivity coeficient depending on the diference of normal displacements on Γ_c, i.e. on the slot between the bodies. Besides, $w^{(l)}$ denotes the restriction of w (considered as defined on the whole domain Ω) to Ω_l, thus $[u] = u^{(2)} - u^{(1)}$ is well defined on Γ_c.

Furthermore, we will deal with the thermo-visco-elastic system, cf. e.g. [1,2]. Using the summation convention, we consider the equilibrium equations:

$$(6) \qquad \rho_l \frac{\partial^2}{\partial t^2} u_i - \frac{\partial \tau_{ij}(u,\theta)}{\partial x_j} = f_i \quad \text{on } Q_l,\ l = 1,2,\ i = 1,2,3,$$

where $f = (f_1, f_2, f_3)$ is a loading force, ρ_l are mass densities, and τ_{ij} are the stress tensor components, having stationary and viscous parts:

$$(7) \qquad \tau_{ij}(u,\theta) = \tau_{ij}^s(u,\theta) + \tau_{ij}^v\left(\frac{\partial u}{\partial t}\right),$$

with

$$(8) \qquad \tau_{ij}^s(u,\theta) = \lambda_l(\theta)\nabla u \delta_{ij} + 2\mu_l(\theta)e_{ij}(u) + (3\lambda_l(\theta) + 2\mu_l(\theta))\gamma_l(\theta)\delta_{ij} \quad \text{on } Q_l,$$

and

$$(9) \qquad \tau_{ij}^v\left(\frac{\partial u}{\partial t}\right) = \lambda_l^v \nabla\left(\frac{\partial u}{\partial t}\right)\delta_{ij} + 2\mu_l^v e_{ij}\left(\frac{\partial u}{\partial t}\right) \quad \text{on } Q_l,\ l = 1,2,$$

where $u = (u_1, u_2, u_3)$ is the displacement vector, $e_{ij}(u) = \frac{1}{2}\left(\frac{\partial u_i}{\partial x_j} + \frac{\partial u_j}{\partial x_i}\right)$ are the small strain tensor components, δ_{ij} is the Kronecker symbol, stationar Lamé coefficients λ_l, μ_l are temperature dependent, while the viscous coefficients λ_l^v, μ_l^v are constants, and finally γ_l are temperature dilatations of the material – of course, dependent on temperature.

On the contact boundary Γ_c we consider the standard Signorini conditions without friction:

$$(10) \qquad [u]\nu \leq 0 \quad \text{on } \Sigma_c,$$

$$(11) \qquad \tau_{ij}(u^{(l)}, \theta^{(l)})\nu_i\nu_j \leq 0 \quad \text{on } \Sigma_c,\ l = 1,2,$$

and the complementarity condition

(12) $$[u]\nu(\tau_{ij}(u^{(l)},\theta^{(l)})\nu_i\nu_j) = 0 \quad \text{na } \Sigma_c, \quad l = 1,2.$$

Of course, also the equilibrium conditions on the contact must be considered:

(13) $$\tau_{ij}(u^{(1)},\theta^{(1)})\nu_j + \tau_{ij}(u^{(2)},\theta^{(2)})\nu_j = 0 \quad \text{on } \Sigma_c \quad i = 1,2,3,$$

$$\tau_{kj}(u^{(1)},\theta^{(1)})\nu_j - \tau_{ij}(u^{(1)},\theta^{(1)})\nu_i\nu_j\nu_k = 0 \quad \text{on } \Sigma_c \quad k = 1,2,3.$$

On Γ_l we consider, for simplicity, the Dirichlet boundary conditions:

(14) $$u = 0 \quad \text{on } \Sigma_l, \quad l = 1,2.$$

It remains to complete the system by the initial conditions for the displacement and velocity:

(15) $$u(x,0) = u^0(x), \quad \frac{\partial u}{\partial t}(x,0) = \dot{u}^0(x) \quad \text{for } x \in \Omega = \Omega_1 \cup \Omega_2 \cup \Gamma_c .$$

2. The results in brief.

Because of lack of space we are forced only to outline the results very briefly, referring to [6]. The weak solution of the problem (1)–(15) can be defined by using once the Green formula in space and once the per-partes integration in time both for the heat transfer problem (cf. also [5]) and for the visco-elastic system. Besides, by means of the standard Rothe method for semidiscretization in time (see e.g. [3]) and the standard penalty function method to replace the variational inequality by an equation, we can define an approximate solution denoted by $(w_{\eta\varepsilon}, u_{\eta\varepsilon})$, where $\eta > 0$ is the time step for the Rothe method and $\varepsilon > 0$ is the parameter for the penalty function method. For the argument of $r(.)$, we have used the diference of normal displacements from the previous time level (not from a current one), thus the time discretization is not fully implicit, but only semi-implicit, which simplifies the proof of the existence of the approximate solution. Besides, we have not used numerical integration in the penalty term, which simplifies the apriori estimates.

Under quite natural assumptions on the data (e.g. $r : \mathbb{R} \to \mathbb{R}$ continuous and greater than a positive constant, $\gamma_l, \lambda_l, \mu_l$ Lipschitz continuous and nonnegative, $\mu_l^v > 0$, $\lambda_l^v \geq 0$, etc., as for α_l, β_l and g_l we refer to [5]), we can prove (essentially by the technique from [5], i.e. a comparison and the Schauder fixed point theorems) that, for every $\eta, \varepsilon > 0$, there is at least one approximate solution $(w_{\eta\varepsilon}, u_{\eta\varepsilon})$ which fulfils the following apriori estimates:

$$(16) \qquad \|w_{\eta\varepsilon}\|_{L^\infty(Q)} \leq C,$$

$$(17) \qquad \|\alpha(w_{\eta\varepsilon})\|_{L^2(0,T;\mathcal{H}^1(\Omega))} \leq C,$$

$$(18) \qquad \left\|\frac{\partial}{\partial t} w_{\eta\varepsilon}^I\right\|_{L^2(0,T;\mathcal{H}^1(\Omega)^*)} \leq C,$$

$$(19) \qquad \left\|\frac{\partial}{\partial t} u_{\eta\varepsilon}^I\right\|_{L^2(0,T;\mathcal{H}^1(\Omega)^3) \cap L^\infty(0,T;L^2(\Omega)^3)} \leq C,$$

$$(20) \qquad \|\dot{u}_{\eta\varepsilon}\|_{H^1(0,T;(H_0^1(\Omega)^*)^3)} \leq C,$$

$$(21) \qquad \|\dot{u}_{\eta\varepsilon}\|_{H^1(0,T;(\mathcal{H}_0^1(\Omega)^*)^3)} \leq C(1 + \varepsilon^{-1/2}),$$

$$(22) \qquad \|[u_{\eta\varepsilon}\nu]^+\|_{L^\infty(0,T;L^2(\Gamma_c))} \leq C\varepsilon^{1/2},$$

where $w_{\eta\varepsilon}^I \in C^0(0,T;L^2(\Omega))$ denotes the piecewise linear interpolation in time of $w_{\eta\varepsilon}$ which is piecewise constant on particular time subintervals, analogously $u_{\eta\varepsilon}^I$ is the linear interpolation of $u_{\eta\varepsilon}$, and $\dot{u}_{\eta\varepsilon} \in C^0(0,T;L^2(\Omega))^3$ is the linear interpolation of the time derivative of $u_{\eta\varepsilon}^I$. Of course, C does not depend on η and ε, and the notation concerning function spaces and their norm is standard, $\mathcal{H}^1(\Omega)$ denotes the space of function whose restrictions on Ω_l belongs to $H^1(\Omega_l)$, i.e. the jumps on Γ_c are admitted.

These apriori estimates enable us to pass to the limit: We take a sequence of the approximate solutions $\{(w_{\eta\varepsilon}, u_{\eta\varepsilon})\}$ such that $\eta,\ \varepsilon \searrow 0$ and $\eta^2/\varepsilon \to 0$. Then there is its subsequence (denoted by the same indices, for simplicty) such that

$$(23) \qquad w_{\eta\varepsilon}^I \to w \quad \text{strongly in } L^2(0,T;\mathcal{H}^1(\Omega)^*) \text{ and}$$

$$(24) \qquad u_{\eta\varepsilon}^I \to u \quad \text{weakly in } H^1(0,T;\mathcal{H}^1(\Omega)^3).$$

Moreover, every (w,u) thus obtained is a weak solution of the problem (1)–(15), and

$$(25) \qquad \alpha(w_{\eta\varepsilon}) \to \alpha(w) \quad \text{weakly in } L^2(0,T;\mathcal{H}^1(\Omega)) \ .$$

The proof of (23) and (25), being only a slight extension of that in [5], exploits particularly the monotonicity technique. The proof of (24) is complicated particularly by the temperature dependence of the Lamé coefficients because we have not got any estimate for the time derivative of the temperature, which forced us to use a compensation-compactness-like technique, partly modifying also the results from [7]. Besides, several other tricks are needed, which makes the full proof rather long.

References

1. G. Duvaut, J.-L. Lions: *Les Inéquations en Mécanique et en Physique*. Dunod, Paris, 1972.

2. J. Jarušek: Contact problems with given time-dependent friction force in linear viscoelasticity. *Com. Math. Univ. Carolinae*. **31** (1990), 257–262.

3. K. Rektorys: *The Method of Discretization in Time and Partial Differential Equations*. D. Reidel Publ., Dordrecht, 1982.

4. J.F. Rodrigues: A unilateral thermoelastic Stefan-type problem. *Portugalie Mathematica* **45** (1988), 91–103.

5. T. Roubíček: The Stefan problem in heterogeneous media. *Annales de l'IHP, Analyse non linéaire* **6** (1989), 481–501.

6. T. Roubíček: A coupled contact problem in nonlinear thermo-visco-elasticity. (submitted)

7. V. Šverák: (to appear)

Tomáš Roubíček
Institute of Information Theory and Automation
Czechoslovak Academy of Sciences
Pod vodárenskou věží 4
CS-182 08 Praha 8
Czechoslovakia

Present address:
Institut für Mathematik der Universität Augsburg
Universitätsstraße 8, W-8900 Augsburg, Germany

P SHI AND M SHILLOR

Recent advances in quasistatic contact problems of thermoelasticity

We survey some recent progress in quasistatic contact problems of thermoelasticity. Since the original work of Fichera [F] who solved Signorini's contact problem of linear elasticity in 1964, contact problems have grown to an spectacular branch of mathematics. However, untill recently there have been only few result of thermoelastic contact from the mathematics point of view because these problems lack coerciveness and monotonicity, so the standard theory of variational inequalities failed to apply. We mention that [D] considered a static contact problem in thermoelasticity; [Rod] and [Rou] considered uncoupled problems with unilateral conditions.

Recently, we have initiated a series investigations of thermoelastic contact within the fully coupled, quasistatic framework. The physical setting for the one dimensional case is depicted in the following Figure.

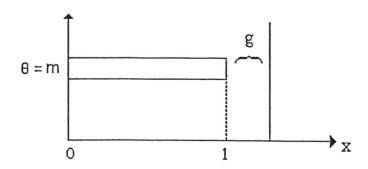

The problem models the quasistatic evolution of a homogeneous rod that is held fixed at $x = 0$. It is free to expand or contract on the other edge and may come into contact with the rigid wall at $x = 1 + g$. Here the interval $(0, 1)$ denotes the initial configuration of the rod and $g \geq 0$ denotes the gap between the reference configuration of the rod and the wall.

In rest of the paper we use $\theta(x, t)$ and $u(x, t)$ to denote the temperature and the displacement. Let $\Omega_T = (0, 1) \times (0, T)$. For definitions of various function spaces we refer the reader to [LSU].

In [GSS] the following problem is considered. Find $\{\theta, u\} \in W_2^{2,1}(\Omega_T) \times W_2^{2,1}(\Omega_T)$ such that

$$\theta_{xx} = -\theta_t + a u_{xt}, \qquad (x, t) \in \Omega_T, \tag{1}$$

$$u_{xx} = a\theta_x, \qquad (x, t) \in \Omega_T, \tag{2}$$

$$\theta = \varphi, \qquad 0 < x < 1, \ t = 0, \tag{3}$$

$$\theta = 1 \qquad x = 0, \ 0 < t < T, \tag{4}$$

$$\theta = 0 \qquad x = 1, \ 0 < t < T, \tag{5}$$

$$u = 0 \qquad x = 0, \ 0 < t < T, \tag{6}$$

and u satisfies Signorini's contact condition

$$u \leq 0, \quad u_x \leq 0, \quad \text{and} \quad u_x(u - g) = 0, \quad x = 1, \ 0 < t < T. \tag{7}$$

The existence of a strong solution $\{\theta, u\} \in W_2^{2,1}(\Omega_T) \times W_2^{2,1}(\Omega_T)$ is proved in in [GSS], provided that $0 < a < 1$.

The problem (1)-(7) turns out to decouple by virtue of the following interesting observation:

$$c = \min\{a, b\} \quad \text{if and only if} \quad c \leq a, \quad c \leq b \quad \text{and} \quad (c - a)(c - b) = 0. \tag{8}$$

The decoupling was first noticed in [SS1]. The temperature θ thereby satisfies the following nonlinear, nonlocal equation

$$(1 + a^2)\theta_t - \theta_{xx} = a\frac{d}{dt} \max\left\{a \int_0^1 \theta(\xi, t)d\xi - g, 0\right\} \quad \text{in } \Omega_T.$$

Once θ is found the position of the edge can be determined by the relation

$$u(1, t) = \min\left\{a \int_0^1 \theta(\xi, t)d\xi, g\right\}, \quad 0 < t < T. \tag{9}$$

This decoupling procedure is the crucial step for the proof of the well-posedness of the problem (1)-(7). It also opened the gate for more quantitative analysis of the one dimensional thermoelastic contact. Along this line, [SS2] considered the radiation condition for the temperature at $x = 1$. The existence and uniqueness of a the strong solution is proved for the problem

$$(1 + a^2)\theta_t - \theta_{xx} = a\frac{d}{dt} \max\left\{a \int_0^1 \theta(\xi, t)d\xi - g, 0\right\} \quad \text{in } \Omega_T, \tag{10}$$

$$\theta(0, t) = m, \qquad\qquad\qquad\qquad 0 < t < T, \tag{11}$$

$$\theta(x, 0) = \varphi(x), \qquad\qquad\qquad\qquad 0 < x < 1, \tag{12}$$

$$-\theta_x(1, t) = k\theta(1, t), \qquad\qquad\qquad 0 < t < T, \tag{13}$$

where $k > 0$ is a constant, and a is an arbitrary positive number, thus improving the restriction that $0 < a < 1$ in [GSS].

The behavior of of the heat conductivity k when the right edge of the bar is sufficiently close to the wall led Barber [B] to model the limiting case, which he called the imperfect thermal condition, for which k is assumed as a graph. Based on the work of [B], the following reformulation of Barber's heat exchange condition is postulated in [ASSW].

$$k = k(r), \quad \text{where} \quad r = \begin{cases} \text{gap size} & \text{if there is no contact;} \\ \text{contact pressure} & \text{if there is contact.} \end{cases} \tag{14}$$

Surprisingly, this physical assumption can be written in the elegant form

$$k = k\left(a \int_0^1 \theta(\xi, t)d\xi - g\right) \tag{15}$$

with the aid of (8). Substituting (15) into (13), it follows that

$$-\theta_x(1,t) \in k \left(a \int_0^1 \theta(\xi,t)d\xi - g \right) \cdot \theta(1,t),$$ (16)

where $k(\cdot)$ is a graph, possibly discontinuous at zero.

The existence of a weak solution in the space $W_2^{1,1/2}(\Omega_T)$ to the problem (10)-(12) and (16) is proved in [ASSW], proveided that $0 < a < 1$. The proof requires rather delicate estimates, for which we refer the reader to the original work.

The first numerical scheme with proven convergence for problem (10)-(13) is constructed in [SSZ]. The numerical solution has shown interesting oscillatory behavior. Using the Dirichlet condition (4)-(5) with appropriate choice of the initial temperature φ, the right edge of the bar first contracts, and then expands and gets in touch with the wall. After staying in contact with the wall for a while, it gradually elvoves towards the steady state. [CE] considered a finite element approximation to (1)-(7) and derived error estimates. They also proved the existence of a forced periodic solution to the problem.

The n-dimensional quasistatic contact problem in thermoelasticity has been considered in [SS3]. The existence of the so called strong-weak solution is proved for the problem by compactness argument together with truncation.

Since the problem is relatively new, many basic questions remain unsolved. For example, the uniqueness of the weak solution to (10)-(12) and (16) is open. The L^p theory for problem (10)-(13) is completely unexplored. We remark that, if k in (16) is assumed monotone with sufficiently large slope, it is then easy to show that there are three different solutions for the steady state. This suggests different initial conditions may lead to differently steady states. It would be interesting to know if for some particular class of initial conditions the solution to the evolution problem would approach no steady state and thereby form autonomous oscillations.

REFERENCES

[ASSW] K. Andrews, P. Shi, M. Shillor and S. Wright, "Thermoelastic Contact with Barber's Heat Exchange Condition", Preprint.

[B] J. R. Barber, "Contact problems involving a cooled punch", *J. Elasticity,* 8 (1978), 409-423.

[CD] M. Comninou and J. Dundurs , "On the Barber boundary conditions for thermoelastic contact ", *J. Appl. Mech.*, **46** (1979), 849-853.

[CE] M.I.M. Copetti and C.M. Elloitt, "A One Dimensional Quasistatic Contact Problem in Linear Thermoelasticity", to appear in *Euro. J. Appl. Math.*.

[D] G. Duvaut, "Free boundary problems connected with thermoelasticity and unilateral contact", *Free Boundary Problems* Vol II, Roma 1980.

[F] G. Fichera, "Boundary value problems of elasticity with unilateral constraints", in *Handbuch der Physic* VIa/2, Springer, Berlin, 1972, 391-424.

[GSS] R. P. Gilbert, P. Shi and M. Shillor, "A quasistatic contact problem in linear thermoelasticity ", *Rediconti di Mat.*, **10** (1990), 785-808.

[LSU] O. A. Ladyzenskaja, V. A. Solonnikov and N. N. Uralceva, *Linear and Quasilinear Equations of Parabolic Type*, American Mathematical Society, Providence, Rhode Island, 1968.

[Rod] J. F. Rodrigues, " A unilateral thermolastic Stefan-type problem", *Portug. Math.*, **45** (1988), 91-103.

[Rou] T. Roubiček, "A coupled contact problem in nonlinear thermo-visco-elasticity", preprint.

[SS1] P. Shi and M. Shillor, "Uniqueness and stability of the solution to a thermoelastic contact problem", *Euro. J. Appl. Math*, **1** (1990), 371-387.

[SS2]   P. Shi and M. Shillor, "A quasistatic contact problem in thermoelasticity with a radiation condition for the temperature", to appear in *JMAA*.

[SS3]   P. Shi and M. Shillor, "Existence of a solution to the n-dimensional problem of thermoelastic contact", preprint.

[SSZ]   P. Shi, M. Shillor and X. L. Zou, "Numerical solutions to one dimensional problems of thermoelastic contact", *Comput. Math. Appl.*, **27** (1991), 65-78.

PETER SHI AND MEIR SHILLOR

Depatment of Mathematical Sciences
Oakland University
Rochester, MI 48309-4401 U.S.A.

M SHILLOR AND P SHI

On the regularity of solutions to problems with friction

We present a new method for the study of regularity of solutions to contact problems with friction. Such problems can be formulated as variational inequalities with friction type functionals. These functionals are sublinear, nondifferentiable and, what is crucial to our method, include H_0^1 in their kernel. The idea is based on an appropriate characterization of subgradients of the friction functional. Then the question of the regularity of a solution to the variational inequality is reduced to that of a solution to an appropriate elliptic boundary value problem.

The regularity of solutions to problems of the form (1), below, was considered in Lions [L], Cocu and Radoslovescu [CR], Kato [K], Gastaldi and Gilardi [GG] and Gastaldi [G]. They all used the standard method of translation parallel to the boundary (Nirenberg [Ni]). Our method is completely different.

First we present the main ideas in an abstract form. Then we discuss shortly applications to the membrane problem with friction and to the frictional contact with normal compliance. Full details of the proofs and the examples are given in [SS].

Let $\Omega \subset R^n$ be a bounded domain with smooth boundary $\partial\Omega$. Let V be a Hilbert space of functions on Ω, with continuous imbedding $V \hookrightarrow H^1(\Omega)$. Let V' denote the dual. Assume that $a : V \times V \to R$ is a continuous bilinear form and let $j : H^1(\Omega) \to R$ be the sublinear functional $j(v) = \langle g, |v| \rangle$, for some $g \in H^{-\frac{1}{2}}(\partial\Omega)$, where $\langle \cdot, \cdot \rangle$ is the duality pairing between $H^{\frac{1}{2}}(\partial\Omega)$ and $H^{-\frac{1}{2}}(\partial\Omega)$. Consider the variational inequality

$$u \in V, \ a(u, v - u) + j(v) - j(u) \geq (F, v - u), \qquad \forall \, v \in V. \qquad (1)$$

Here $F \in V$ and $(\cdot, \cdot)$ is the inner product in $H^1(\Omega)$.

We define an operator $A : V \to V$ by

$$(Au, v) = a(u, v), \qquad \forall \, v \in V.$$

Then u is a solution to (1) iff

$$j(v) - j(u) \geq (-Au + F, \ v - u), \qquad \forall \, v \in V.$$

That is to say $-Au + F \in \partial j(u)$, where ∂j denotes the subdifferential of j. It follows that there exists a subgradient $u^* \in \partial j(u)$ such that

$$-Au + F = u^* \quad \text{in } H^1(\Omega). \qquad (2)$$

Then the regularity of u can be obtained from the regularity of u^* and of the regularity of solutions to (2).

Assume that $g \in L^p(\partial\Omega)$, for some $1 < p < \infty$. Then it is shown that if $w^* \in \partial j(u)$ then $-\Delta w^* + w^* = 0$, in $H^1(\Omega)$ and $\exists \varphi \in L^p(\partial\Omega)$ such that $\partial w^* / \partial n = \varphi$, a.e. on $\partial\Omega$. If F is smooth we obtain from (2) that $(-Au + F, v) = (u^*, v)$, $\forall \, v \in V$. Let f be such that $(F, v) = \int_\Omega fv dx$, $\forall \, v \in V$. Then we obtain that

$$-a(u, v) + \int_\Omega fv dx = \int_\Omega \nabla u^* \nabla v dx + \int_\Omega u^* v dx, \qquad \forall v \in V. \qquad (3)$$

Next, to simplify the exposition, assume that

$$a(u, v) = \int_\Omega \nabla u \, \nabla v dx, \qquad \forall v \in V, \qquad (4)$$

69

i.e. the bilinear form associated with the Laplacian. Applying Greens formula and (4) to (3) gives

$$\Delta u + f = -\Delta u^* + u^* \ \text{in} \ \Omega, \qquad \frac{\partial u}{\partial n} = \frac{\partial u^*}{\partial n} \ \text{on} \ \partial\Omega. \qquad (5)$$

But $u^* \in \partial j(u)$ so $-\Delta u^* + u^* = 0$ in Ω and $\partial u^* \partial n = \varphi$ a.e. on $\partial\Omega$. Thus

$$-\Delta u = f \qquad \text{in} \ \Omega, \qquad \frac{\partial u}{\partial n} = \varphi \qquad \text{on} \ \ \partial\Omega. \qquad (6)$$

We conclude that u, being a solution to (1), is also a solution to (6) and its regularity follows from the regularity results for (6).

Above we used V in the formulation (1). Similar results are obtained when V is replaced by a closed, convex, nonempty set $K \subset V$.

Membrane with Friction. The problem was considered in [L] and its regularity in [GG]. Let $\Omega \subset R^2$ be a bounded domain, representing the vertical projection of the membrane. Vertical forces of surface density f act on the membrane. It is subject to contact with friction on Γ_C, which is a part of $\partial\Omega$, with friction bound g. It is held fixed on the rest of the boundary, Γ_D (possibly $\Gamma_D = \emptyset$). The classical formulation of the problem ([DL] or [L]) is to find $u : \Omega \to R$ such that

$$|\frac{\partial u}{\partial n}| \leq g \qquad \text{on} \ \Gamma_C \ \text{and}$$

$$|\frac{\partial u}{\partial n}| < g \qquad \Rightarrow \qquad u = 0, \qquad\qquad\qquad (7)$$

$$|\frac{\partial u}{\partial n}| = g \qquad \Rightarrow \qquad u = -\lambda\frac{\partial u}{\partial n} \quad \text{for some} \ \lambda \geq 0.$$

Here u is the vertical displacement of the membrane, **n** the unit outward normal to $\partial\Omega$ and $\partial/\partial n$ is the normal derivative. Condition (7) is the friction condition. The region in Γ_C where $|\partial u/\partial n| < g$ is called the *stick region* and the region in Γ_C where $|\partial u/\partial n| = g$ is the *slip region*. The investigation of these regions as well

70

as the boundary between them, which is a free boundary, depends crucially on the regularity of solutions to the problem.

The problem can be set as a variational inequality (1) with a as in (4).

Theorem 1.([SS]) *Assume that $f \in L^p(\Omega)$, $\partial\Omega \in C^{1,1}$ and $g \in L^p(\Gamma_C)$, $2 \leq p < \infty$. If $u \in H^1(\Omega)$ is a solution to (1), then*

$$u \in W^{s,p}(\Omega) \quad \text{if } \Gamma_D = \emptyset, \quad s < 1 + \frac{1}{p} \text{ if } p > 2, \quad s = \frac{3}{2} \text{ if } p = 2;$$

$$u \in W^{1,p}(\Omega) \quad \text{if } \Gamma_D \neq \emptyset, \text{ for } 2 \leq p < 4.$$

Frictional contact problem with normal compliance. Problems of this type were considered in [OM], [M], [KO], [KMS1, 2, 3, 4, 5], [EMS] and [SS]. The derivation and some background can be found in [OM] or [KMS1]. The variational inequality formulation is as follows. Let $V = \{\mathbf{v} \in \mathbf{H}^1(\Omega); \ \mathbf{v} = 0 \text{ on } \Gamma_D\}$, where $\mathbf{H}^1(\Omega) = H^1(\Omega; R^2)$, $\Omega \subset R^2$ and $\partial\Omega = \overline{\Gamma}_C \cup \overline{\Gamma}_D$. Let $a(\mathbf{u}, \mathbf{v})$ be the usual bilinear form in linear elasticity, let $j_N : \mathbf{H}^1(\Omega) \times \mathbf{H}^1(\Omega) \to R$,

$$j_N(\mathbf{u}, \mathbf{v}) = \int_{\Gamma_C} C_N[(u_N - g_a)_+]^{m_N} v_N \, ds, \tag{8}$$

be the *normal compliance form* and $j_T : \mathbf{H}^1(\Omega) \times \mathbf{H}^1(\Omega) \to R$,

$$j_T(\mathbf{u}, \mathbf{v}) = \int_{\Gamma_C} C_T[(u_N - g_a)_+]^{m_T} |v_T| ds,$$

be the *friction functional*. Here $u_N = \mathbf{u}\mathbf{n}$ and $\mathbf{u}_T = \mathbf{u} - u_N\mathbf{n}$ are the normal and tangential components of $\mathbf{u}$ on Γ_C. The contact problem may be written as

$$\mathbf{u} \in V, a(\mathbf{u}, \mathbf{v} - \mathbf{u}) + j_N(\mathbf{u}, \mathbf{v} - \mathbf{u}) + j_T(\mathbf{u}, \mathbf{v}) - j_T(\mathbf{u}, \mathbf{u}) \geq F(\mathbf{v} - \mathbf{u}), \ \forall \ \mathbf{v} \in V. \tag{9}$$

Theorem 2. ([SS]) *Assume that $\Omega \subset \mathbf{R}^2$ is a bounded domain with $\partial\Omega$ in $C^{1,1}$. Let $f \in L^\infty(\Omega), C_T, C_N \in W^{1,\infty}(\Gamma_C), 1 \leq m_N, m_T < \infty, g_a \in C^1(\overline{\Gamma}_C)$. If $\mathbf{u}$ is a* *solution to (9) then*

$$\mathbf{u} \in W^{1,p(\Omega)}, \qquad 2 \leq p < 4. \tag{10}$$

Similar regularity was obtained in [KMS4] using this method (see also [KMS5]).

REFERENCES

[CR]   Cocu, M. and Radoslovescu, A. "Regularity properties for the solutions of a class of variational inequalities", *Nonlin. Anal. TMA* **11** (2) (1987), 221-230.

[DL]   Duvaut, G. and Lions, J. L. *Inequalities in Mechanics and Physics* Springer-Verlag, New York, 1976.

[EMS]   Elliott, C.M., Mikelić, A. and Shillor, M. "Constrained anisotropic elastic materials in unilateral contact with or without friction", submitted for publication.

[G]   Gastaldi, F. "Remarks on noncoercive contact problems with friction in elastostatics" *IAN* publication 649, 1988.

[GG]   Gastaldi, F. and Gilardi, G. "A class of noncoercive variational inequalities", to appear.

[K]   Kato, Y. "Signorini's problem with friction in linear elasticity", *Japan J. Appl. Math.* **4** (1987), 237-268.

[KO]   Kikuchi, N. and Oden, J.T. *Contact Problems in Elasticity: A Study of Variational Inequalities and Finite Element Methods.* SIAM, Philadelphia 1988.

[KMS1] Klarbring, A., Mikelić, A. and Shillor, M. "Frictional contact problems with normal compliance", *Int. J. Engng. Sci.* **26** (8) (1988), 811-832.

[KMS2] Klarbring, A., Mikelić, A. and Shillor, M. "On friction problems with normal compliance", *Nonlin. Anal. TMA* **13** (8) (1989), 935-955.

[KMS3] Klarbring, A., Mikelić, A. and Shillor, M. "Duality applied to contact problems with friction", submitted for publication.

[KMS4] Klarbring, A., Mikelić, A. and Shillor, M. "Mathematical analysis of a problem of a rigid punch with friction", in preparation.

[KMS5] Klarbring, A., Mikelić, A. and Shillor, M. "Mathematical analysis of a problem of a rigid punch with friction", this volume.

[L]   Lions, J. L. "Partial differential inequalities", *Russian Math. Surv.* 27 (2) (1972), 91-161.

[M]   Martins, J.A.C. *Dynamic Frictional Contact Problems Involving Metalic Bodies.*
Ph.D. thesis, University of Texas at Austin, 1986.

[Ni]   Nirenberg, L. "Remarks on strongly elliptic partial differential equations", *Comm. Pure Appl. Math.* 8 (1955), 648-674.

[OM]   Oden, J.T. and Martins, J.A.C. "Models and computational methods for dynamic friction phenomena", *Comput. Meth. Appl. Mech. Engrg.* 52(1985), 527-634.

[SS]   Shi, P. and Shillor, M. "Regularity results for problems with friction," submitted.

M. Shillor and P. Shi
Department of Mathematical Sciences, Oakland University
Rochester, MI 48309. USA

D R WESTBROOK
The folding of bands of fibres on dryers

The problem is to model the folding of a broad band of fibres which is fed continuously from a faster to a slower rotating cylindrical drum. The drums are perforated and the band of fibres is held on the drums by air suction through the perforations. Fixed baffles inside each drum are used to ensure that fibres are fed from one drum to the other.

Because the second drum is rotating at about 85–90% as fast as the first drum the band is overfed as it moves from the first to the second drum and some folding is unavoidable. If the baffles are placed a little before the common diameter of the drums it is observed that there are even folds across the band and that they form a regular pattern (see Figures 1 and 2).

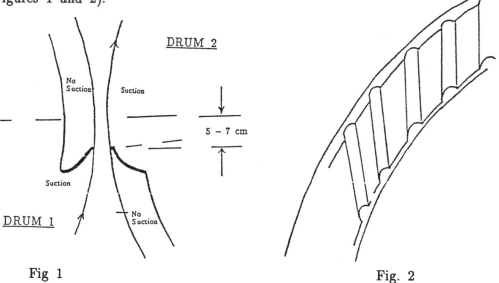

Fig 1

Fig. 2

To obtain a simple model the following assumptions are made.

The band is considered as a thin plate under plane stress. Because the density is small and the velocity is almost constant along the band inertial effects are neglected. Friction between the drums and the band is also ignored. It is also assumed that stretching tangential to the plate is small. Because of the baffles and the observed fact that the band leaves the first drum after the baffle it is assumed that the suction acts towards the second drum. For simplicity it was taken as constant and normal to the band.

The differential equations are essentially those of the elastica. If ψ is the angle between the tangent to the elastica and a line perpendicular to the common diameter of the drums, n is the normal stress tangential to the elastica, q is the shear stress and m the bending movement. Then the equations of equilibrium are

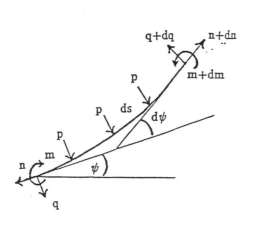

fig. 3

$$\frac{dn}{ds} - q\frac{d\psi}{ds} = 0$$

$$\frac{dq}{ds} + n\frac{d\psi}{ds} - p = 0$$

$$\frac{dm}{ds} + q = 0$$

where s is the arc length along the curve. In addition to these equations one has a constitutive equation which is assumed to be the Euler Bernoulli beam law (this also applies to the plane stress plate situation) $m = EI\frac{d\psi}{ds}$. Since boundary conditions of position must be included in the formulation we also have the equations $\frac{dx}{ds} = \cos\psi$, $\frac{dy}{ds} = \sin\psi$ with x measured vertically down. For convenience I choose the origin of the xy system at the centre of the second drum with the centre of the first drum at (0,2R+2h) where R is the common radius of the drums and 2h the gap between the drums. I also use another pair of cartesian coordinates (ξ, η) with the same origin but with the axes rotated through an angle ψ so that $\xi = x\cos\psi + \eta\sin\psi$, $\eta = -x\sin\psi + y\cos\psi$. ξ and η satisfy the equations $\frac{d\xi}{ds} = 1 + \eta\frac{d\psi}{ds}$, $\frac{d\eta}{ds} = -\xi\frac{d\psi}{ds}$. It is probable that a linearized version of the equations under the assumption that ψ and q are small would give reasonable results. This would lead to the standard linear beam equations with n = constant = $-\lambda^2$, x = s and m = $EI\frac{d^2w}{dx^2}$, $q = -EI\frac{d^3w}{dx^3}$, $EI\frac{d^4w}{dx^4} + \lambda\frac{d^2w}{dx^2} + p = 0$. At points of contact on the second drum $\xi = 0$, $\eta = R$ and $\frac{d\psi}{ds} = -\frac{1}{R}$. On the first drum if the angle of contact is ψ_1 then $\xi = 2(R + h)\sin\psi_1$ $\eta = 2(R + h)\cos\psi_1 - R$, $\frac{d\psi}{ds} = \frac{1}{R}$. ψ_1 and the angle of contact ψ_2 on the second drum are unknowns and $\frac{d^2\psi}{ds^2} = -\frac{q}{EI}$ need not be zero at these points. Thus n and q are also unknown at the contact points. The remaining condition is the overfeed condition which may be written in the form

$$L - R(\psi_1 + \psi_2) = \text{const} + (w_1 - w_2)Rt.$$

Here L is the length of the band between the drums, ψ_1 and ψ_2 are the values of the angle ψ at the contact points on the drums and w_1, w_2 are the angular velocities of the drums.

The problem then has a free boundary on each drum with an unknown length of fibre between them. The solution of the sixth order system must satisfy three conditions at s = 0 and two relations at s = L, L being unknown. In addition there is the overfeed condition and the unilateral constraints that the fibre band remains between the drums.

An appealing possible mechanism for the formation of loops would be that as the overfeed is increased part of the band approaches and finally touches the second drum. The mechanism however may be ruled out as follows.

First note the following integrals of the system of equations

$$n = A\cos\psi + B\sin\psi - p\eta \qquad \tfrac{1}{2}EI\psi'^2 + n = \tfrac{1}{2}\frac{EI}{R^2} + n_2$$

$$q = -A\sin\psi + b\cos\psi + p\xi \qquad EI\psi' + q\xi - n\eta - \tfrac{1}{2}p(\xi^2 + y^2) = \frac{EI}{R} - n_2R - \tfrac{1}{2}pR^2$$

A and B are constants and the conditions $\xi = 0$, $\eta = R$, $\psi' = -\frac{1}{R}$ $n = n_2$ at s = 0 have been used. For the band to touch the second drum again at s = s_p where $\xi = 0$, $\eta = R$, $n = n_p$, $\psi = \psi_p$ and $\psi' = \psi'_p$ we must have $\tfrac{1}{2}EI\psi'^2_p + n_p = \tfrac{1}{2}\frac{EI}{R^2} + n_2$, $EI\psi'_p - n_pR = -\frac{EI}{R} - n_2R$. Eliminating n_p and n_2, $\tfrac{1}{2}EI(\psi'_p + \frac{1}{R})^2 = 0$ so that $\psi'_p = -\frac{1}{R}$. For the band to just touch at p however with $\psi'_p = -\frac{1}{R}$ we must also have $\psi''_p = 0$ and hence $q_p = 0$. As a result $n_p = n_2$ and $A\cos\psi_p + B\sin\psi_p = A\cos\psi_2 + B\sin\psi_2$, $q_p = 0$ and $-A\sin\psi_p + B\cos\psi_p = 0$. Thus $\psi_p = \psi_2$, and there is no loop, or $A = B = 0$. If $A = B = 0$ $n = -p\eta$, $q = p\xi$ and therefore $EI\psi' + \tfrac{1}{2}p(\xi^2 + \eta^2) = -\frac{EI}{R} + \tfrac{1}{2}pR^2$. If we write $r^2 = \xi^2 + \eta^2$ and note that $\frac{d}{ds}r^2 = 2\xi$ we have the following equations for ξ, η, ψ.

$$EI\psi' = -\frac{EI}{R} - \tfrac{1}{2}p(r^2 - R^2), \quad \xi' = 1 + \eta\psi', \quad \eta' = -\xi\psi'.$$

Consider now $f(\xi,\eta) = \xi^2 + (\eta - R)^2 + \frac{pR}{4E}(r^2 - R^2)^2$ noting that $f(\xi,\eta) \geq 0$ $f(0,R) = 0$ we see that

$$\frac{d}{ds}f(\xi,\eta) \quad = 2\xi(1 + \eta\psi') - 2(\eta - R)\xi\psi' + \tfrac{pR}{EI}2\xi(r^2 - R^2)$$

$$= 2\xi[1 + R\psi' + \tfrac{pR}{2EI}(r^2 - R^2)] = 0$$

thus $f(\xi,\eta) \equiv 0$ and $\xi = 0$, $\eta = R$, $r = R$ which of course means that the band lies on the drum.

To hopefully restore the possibility of this looping mechanism a gravitational force in the x direction was added. This leads to the new differential equations

$$\frac{dn}{ds} - q\frac{d\psi}{ds} = -\rho g\cos\psi, \qquad EI\frac{d\psi}{ds} = m$$

$$\frac{dq}{ds} + n\frac{d\psi}{ds} = p + \rho g\sin\psi \qquad \frac{d\xi}{ds} = 1 + \eta\frac{d\psi}{ds}$$

$$\frac{dm}{ds} + q = 0 \qquad \frac{d\eta}{ds} = -\xi\frac{d\psi}{ds}$$

with boundary conditions as before. Several numerical solutions of this boundary value problem were found by shooting methods and the satisfaction of the constraint conditions was checked. Solutions were found with only one point of contact on each drum, solutions with two contact points on the second drum and the band wrapped on the second drum at both ends, and solutions with wrapping on both drums with an additional point of contact on the second drum. In this case the shear stress q may undergo a jump at the additional contact point and it is not a limiting solution of the type previously discussed (unless the jump is zero). The jump must be of the correct sign so that a positive reactive force from the drum results. These results were obtained mostly by specifying n_2 instead of the overfeed.

Once a touching solution had been obtained other solutions in which (a) the overfeed was decreased incrementally and (b) increased incrementally

78

were considered. In the case where the overfeed was decreased the loop flattened and appears to approach the drum. As the overfeed increases the loop gets bigger.

This problem is far from solved and work is continuing. So far the numerical experiments have not produced a mechanism for the formation of the loops. Whether this is a consequence of a lack of understanding of the solutions or of the simplicity of the model is not yet clear.

Solidification I

V ALEXIADES AND J B DRAKE

A weak formulation for phase-change problems with bulk movement due to unequal densities[*]

* This research was sponsored by the Applied Mathematical Sciences Research Program, Office of Energy Research, U.S. Department of Energy under contract DE-AC05-84OR21400 with the Martin Marietta Energy Systems, Inc. AND by The Science Alliance, a Centers of Excellence Program of the State of Tennessee at The University of Tennessee, Knoxville.

ABSTRACT

When the densities of solid and liquid are different, a change of phase induces a volume change, forcing movement of the bulk phases. Conservation of total energy across the interface leads to a modified Stefan Condition, which contains a term cubic in the interfacial speed. We present a weak formulation for the conservation of mass, momentum and energy and study numerically a 1-phase, 1-dimensional example.

1. Introduction

When the densities of solid and liquid are different, a change of phase induces a volume change, forcing movement of the bulk phases. The phase-change process is no longer purely thermal, the enthalpy (heat content) is no longer the total energy, and one must consider conservation of mass and momentum in addition to conservation of energy.

A simple problem illustrates best the issues arising from the density change. As model problem we discuss the case of *volume expansion upon freezing* in the simplest 1-dimensional situation: Consider a liquid at its melt temperature, T_m, occupying $0 \leq x < \infty$, and impose a cold temperature $T_{cold} < T_m$ at $x = 0$. Assuming $\rho_S < \rho_L$, the solid forming near the face $x = 0$ will occupy greater volume, thus causing bulk displacement of the liquid to the right (assuming a rigid wall at $x = 0$).

Note that if the densities were equal, this would be the classical 1-phase Stefan Problem, which (for constant properties and data) admits a similarity solution (the Neumann solution). Our objective is to examine the effect of unequal densities on the formulation of the problem, on the solution and, more importantly, to demonstrate how convection can be naturally incorporated in a weak (fixed-domain-type) formulation of the phase-change problem. The formulation and numerical approach are valid for general phase-change problems with convection, in any number of dimensions (see [6], [4], [5]).

2. Conservation laws and interface conditions

Let us consider phase-change processes for a pure material under the following assumptions:
(i) *constant thermophysical properties* $\rho_S \neq \rho_L$, c_p^S, c_p^L, k_S, k_L, T_m, L; (ii) *no viscous dissipation* (so that the stress is simply the pressure); (iii) *no changes in potential energy* (so that total energy = internal + kinetic). We restrict our attention to *1-dimensional* processes, for simplicity. Thus, each phase is *incompressible* and the volume change, due to the change of density upon change of phase, results in bulk displacement of one of the phases with *uniform* speed (but time-dependent in general) as we shall see shortly. The general conservation laws take the following form (in 1-dimension):

mass conservation: $$\rho_t + (\rho v)_x = 0, \tag{2.1}$$

$$\tag{2.2}$$

momentum conservation: $\qquad\qquad (\rho v)_t + (\rho vv + P)_x = 0$,

energy conservation: $\qquad\qquad (\rho\varepsilon)_t + (\rho\varepsilon v + q + P v)_x = 0$, $\qquad q = -kT_x$. (2.3a)

Here ρ, ρv, $\rho\varepsilon$ are the mass, momentum and total energy per unit volume, P is the (thermodynamic) pressure and q is the *conductive* heat flux $(q = -kT_x)$. The laws may be written in many other forms [2] ; in particular, the energy equation, (2.3a), may be re-written in terms of the *internal energy*, $u = \varepsilon - \frac{1}{2}v^2$, as

$$(\rho u)_t + (\rho u v + q)_x + P v_x = 0 ;\qquad\qquad (2.3b)$$

this is most convenient for the numerical approach (see §5). However, (2.1 - 3a) have the great advantage of being in "divergence form", hence, we can immediately write down the jump conditions for mass, momentum and energy across an interface $x = X(t)$:

$$[\![\,\rho\,]\!]_-^+ \, X'(t) = [\![\,\rho v\,]\!]_-^+ \qquad \text{on} \quad x = X(t) ,\qquad\qquad (2.4)$$

$$[\![\,\rho v\,]\!]_-^+ \, X'(t) = [\![\,\rho v^2 + P\,]\!]_-^+ \qquad \text{on} \quad x = X(t) ,\qquad\qquad (2.5)$$

$$[\![\,\rho\varepsilon\,]\!]_-^+ \, X'(t) = [\![\,\rho\varepsilon v + P v + q\,]\!]_-^+ \qquad \text{on} \quad x = X(t) .\qquad\qquad (2.6)$$

We shall use them to derive the correct interface conditions for the classical formulation.

3. Classical Formulation

The classical approach is to impose the conservation laws *separately* inside each phase, and also impose the jump relations on the interface in order to ensure conservation everywhere.

Inside the liquid or the solid, where $\rho \equiv const.$, (ρ_L and ρ_S respectively), the equations may be simplified considerably. Indeed, the continuity equation, (2.1), implies that the velocity must be *uniform* in each phase, so v_L and v_S may only depend on time. In our example, the face $x = 0$ is fixed, so the uniform v_S in the solid must be identically zero. Then, the momentum equation in the solid implies $P^S(x, t) \equiv constant \equiv P_0^S$ = value at the interface, and we may choose $P_0^S = P_{ref}$ = reference pressure. Also, with $v_S \equiv 0$, (2.4) yields

$$v_L(t) = [1 - (\rho_S / \rho_L)]X'(t) \; > \; 0,\qquad\qquad (3.1)$$

so the liquid *must* move to the right with non-constant speed; the momentum equation in the liquid tells us that $P^L(x,t)$ will be linear in x, namely,

$$P^L(x,t) = -\rho_L \, v'_L(t) \, [x - X(t)] + P_0^L(t) ,\qquad\qquad (3.2)$$

with P_0^L, the value at the interface, found from (2.5) to be

$$P_0^L - P_{ref} = P_0^S - P_{ref} + \rho_L v_L[X' - v_L] = \rho_S[1 - (\rho_S / \rho_L)]X'^2 .\qquad\qquad (3.3)$$

Moreover, $\rho \equiv const.$ implies $du = c_p \, dT$, with c_p = *heat capacity (specific heat) under constant pressure*. Hence, in the solid, the energy equation (2.3b) reduces to the standard heat equation:

$$\rho_S c_p^S \, T_t = k_S T_{xx} \qquad \text{in the } solid ,\qquad\qquad (3.4)$$

while $T \equiv T_m$, in the *liquid*. Let us emphasize that even though the energy equation has reduced to (3.4), the correct jump condition is still that in (2.6), a point easily overlooked (as for example in [3, p.291]). To see what (2.6) really says, we need expressions for the energy in each phase, which can be shown to be (see [1] for details)

$$\rho u = \begin{cases} \rho_L u_{ref} + [(\rho_L / \rho_S) - 1]P_{ref} + \rho_L L + \rho_L c_p^L[T - T_m] & \text{in the } liquid \\ \rho_S u_{ref} \qquad\qquad\qquad\qquad + \rho_S c_p^S[T - T_m] & \text{in the } solid \end{cases}\qquad (3.5)$$

Here, u_{ref} denotes the value of u at the common reference state, chosen as (ρ_S , T_m , P_{ref}), i.e. solid at its melt temperature at pressure P_{ref} (which must be the pressure at which the latent heat of fusion has been measured, typically 1 *atm*). Then, using $\rho\varepsilon = \rho u + \frac{1}{2}\rho v^2$, $v_S \equiv 0$, (3.1), (3.3) and (3.5), we see that the jump condition (2.6) becomes (after some manipulations):

$$\rho_S L X' - \frac{1}{2}\rho_S [1 - (\rho_S/\rho_L)]^2 X'^3 = [\![-kT_x]\!]_S^L = k_S T_x , \qquad x = X(t), \qquad t > 0 . \tag{3.6}$$

Summarizing, the classical formulation of the 1-phase problem with volume expansion upon freezing is to find $T(x,t)$ and $X(t)$ satisfying:

$\qquad$ (3.4) for $0 < x < X(t)$, $t > 0$, $\quad T(0,t) = T_{cold}$, $\quad X(0) = 0$, $\quad T(X(t),t) = T_m$, and (3.6).

The presence of the X'^3 - term does not allow a similarity solution. By dropping this *negative* term we slow down the process (increase the latent heat), so we expect that the explicit (Neumann) solution of the resulting 1-phase Stefan Problem will *underestimate* the true interface.

4. Weak Formulation

$\qquad$ The *internal energy* per unit volume, $\rho u \equiv \rho\varepsilon - \frac{1}{2}\rho v^2 \equiv \rho h - P$ (with $h =$ enthalpy), is the most convenient energy potential for our purposes. We need an equation of state, relating ρu with the state variables (ρ, T, P), and valid in any phase: solid, liquid, or mushy (mixture of solid and liquid). In each pure phase it is given by (3.5). In the mushy region (interface), where liquid and solid coexist at temperature T_m, the density, energy, pressure, and momentum undergo changes from their "liquidus" values to their "solidus" values. Note that the jump of ρu at $T = T_m$ is:

$(\rho_L - \rho_S)[u_{ref} + P_{ref}/\rho_S] + \rho_L L \equiv (\rho_L - \rho_S) h_{ref} + \rho_L L$; by choosing the enthalpy at the reference state as $h_{ref} := 0$, the jump is simply the latent heat $\rho_L L$. A very convenient, and physically meaningful, *phase-indicator* is $\lambda(x,t)$, the **volume fraction of liquid** present at x at time t. Any volumetric quantity can be expressed, globally, as a convex combination of its liquid and solid values, in particular,

$$\rho = \lambda\rho_L + (1-\lambda)\rho_S \tag{4.1}$$

$$\rho u = \lambda(\rho u)^L + (1-\lambda)(\rho u)^S = \rho u_{ref} + \lambda[(\rho_L/\rho_S) - 1]P_{ref} + \lambda\rho_L L + (\overline{\rho c_p})[T - T_m]$$

$$\equiv [\rho h_{ref} - P_{ref}] + \lambda\rho_L L + (\overline{\rho c_p})[T - T_m], \tag{4.2}$$

where

$$(\overline{\rho c_p}) := \lambda\rho_L c_p^L + (1-\lambda)\rho_S c_p^S . \tag{4.3}$$

Choosing $h_{ref} := 0$ and $P_{ref} := 0$ for convenience, we have the desired equation of state:

$$\rho u = \lambda\rho_L L + (\overline{\rho c_p})[T - T_m] , \tag{4.4}$$

and the phases may be described by the following alternatives:

$\qquad$ **solid:** $\quad \rho = \rho_S, \quad T < T_m \qquad$ or $\quad \lambda = 0 \qquad$ or $\quad \rho u < 0$

$\qquad$ **mushy:** $\quad \rho = \lambda\rho_L + (1-\lambda)\rho_S, \quad T = T_m \quad$ or $\quad 0 < \lambda < 1 \quad$ or $\quad 0 < \rho u < \rho_L L$

$\qquad$ **liquid:** $\quad \rho = \rho_L, \quad T > T_m \qquad$ or $\quad \lambda = 1 \qquad$ or $\quad \rho_L L < \rho u$.

Note that, given the value of ρu, we can find the liquid fraction from

$$\lambda = \max\{ 0, \ \min\{ 1, \rho u/\rho_L L \} \} , \tag{4.5}$$

and then find the temperature from

$$T = T_m + \frac{\rho u - \lambda\rho_L L}{\lambda\rho_L c_p^L + (1-\lambda)\rho_S c_p^S} , \tag{4.6}$$

in all possible cases. Note that for $\rho_S = \rho_L$, (4.4 - 6) reduce to the relations appropriate for the standard Stefan Problem.

$\qquad$ Summarizing, the weak formulation consists of: the conservation laws (2.1), (2.2), (2.3b), valid globally in an appropriate weak sense, with appropriate initial and boundary conditions, (for our example:

84

$\rho(x, 0) = \rho_L$, $v(x, 0) = 0$, $T(x, 0) = T_m$, $v(0, t) = 0$, $T(0, t) = T_{cold} < T_m$) and the equation of state (4.4). The PDEs determine ρ, ρv and ρu, and then (4.4 - 6) determine the temperature T. In the fluid dynamics literature, this type of formulation is known as "volume of fluid" approach. It is easy to see that in the case $\rho_S = \rho_L$ the formulation collapses to the standard enthalpy formulation of the classical Stefan Problem.

5. Numerical Method

In this section we propose a numerical method for the approximate solution of the above equations which does not explicitly track the front. Such methods have proved very useful for engineering calculations involving phase changes and we would like to extend their application to the case where density changes are important.

To specify the numerical algorithm, consider a mesh of control volumes $[x_{i-1/2}, x_{i+1/2}]$, with the faces $x_{i+1/2}$ taken midway between the internal nodes x_i and x_{i+1}. The "half" points will be the locations of the velocity and we adopt the indexing convention that $v_i \equiv v(x_{i-1/2})$. This mesh is sometimes referred to as a MAC mesh, or simply a staggered grid. We employ a "control-volume" discretization of the weak equations, (2.1), (2.2), (2.3b), in space (which amounts to central differences for all space derivatives), and a predictor-corrector method in time. Since the spatial discretization is standard we will describe the time stepping scheme using the continuous derivative notation.

The time advance begins with an explicit prediction of the new energy. The predicted quantities will be denoted by the superscript *. From the known values at time t_n, we predict ρu^* and ρv^* by

$$\rho u^* = \rho u - \Delta t \, (\rho u v + q)_x \,, \qquad \rho v^* = \rho v - \Delta t \, (\rho v v)_x \,, \qquad (5.1)$$

and estimate the liquid fractions from (4.5). A prediction of the density change over the time step can be made using the computed liquid fractions and (4.1), thus estimating the term ρ_t^{n+1}. To balance mass, the velocities are corrected using a pressure equation derived from the momentum equation (2.2) and the density equation (2.1). This pressure equation is

$$P_{xx} = (\, (\rho v^*)_x + \rho_t^{n+1} \,) \, / \, \Delta t. \qquad (5.2)$$

Boundary conditions on this elliptic equation are of Neumann type: at the left boundary $x=0$ there is no outflow, so $P_x=0$; the right boundary, say at $x=1.0$, is an outflow boundary and the appropriate condition is derived from the velocity correction equation:

$$\rho v^{n+1} = \rho v^* - \Delta t \, P_x \,, \qquad (5.3)$$

which gives the outflow pressure boundary condition: $P_x = (\rho v^* - \rho v^{n+1}) \, / \, \Delta t$ at $x=1.0$. With the pressure and the velocity computed from (5.2 - 3) at the new time level, the pressure work term is computed to correct the energy via

$$\rho u^{n+1} = \rho u^* - \Delta t \, P^{n+1} v_x^{n+1} \,. \qquad (5.4)$$

The corrector equations (5.2-4) are iterated until energy converges, thus bringing the mass, velocity and energy at the new time level into balance. Note that, in addition to the product term in equation (5.4), another nonlinearity appears at the outflow boundary condition for pressure at $x=1.0$. A simple fixed point iteration converges quickly. We found that more iterations are required for larger Stefan numbers and larger density changes, which is to be expected since the iteration is the correction for the pressure work contribution to the internal energy. The corrector iteration may fail to converge if large velocities and pressures are encountered. Study of this iteration and development of a more robust numerical technique would greatly enhance the algorithm's usefulness.

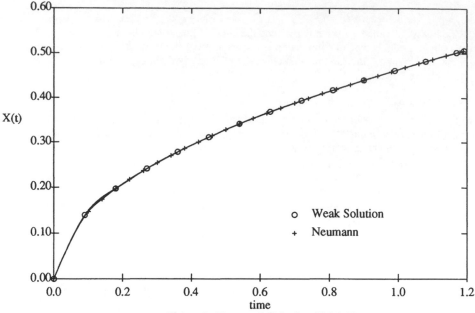

Figure 1: Exact and Calculated Melt Fronts

6.   Calculated Results

The cubic term in the modified Stefan condition (3.6) precludes analytic treatment. By dropping the term, the Neumann solution for the front location can be computed and compared with the numerical solution of the weak formulation. As an example, let $\rho_S = 0.9$, $\rho_L = 1.0$, $T_m = 0$, $L = 10$, $T_{cold} = -1$. **Figure 1** shows the comparison of the Neumann solution, the weak solution with a single density and with two differing densities. Taking $\rho_L = \rho_S$ eliminates the volume expansion and the computed solution accurately reproduces the expected Neumann solution. The weak solution with differing densities increases the speed of the front since part of the energy is converted to kinetic energy. Thus this curve lies above the Neumann solution. The change is very small, however, since the kinetic energy is several orders of magnitude smaller than the latent heat. The curves are indistinguishable graphically. **Figure 2** shows the velocity time-history of a point (x=0.4875) in the middle of the mesh. The velocity shows a series of erratic jumps, then damped oscillations as the front passes from one mass cell to the next. A scalloping curve is quite familiar for front histories computed with the enthalpy method. As the front changes cells, the heat flux to the mushy cell jumps. The rate of freezing suffers a proportionate jump and this is directly related to the velocity. The velocity drops to zero at t=1.1 and remains zero indicating that the material is frozen at that time. Examination of the profiles for velocity and pressure at a given time show a uniform velocity in the liquid (to 5 decimal places) and a linear pressure profile (**Figure 3**). The pressure in the solid is computed to be zero. The pressure jump across the melting cell or cells indicates the ability of the scheme to approximate a discontinuous pressure as a solution of the elliptic pressure equation.

References

[1]   Alexiades, V. and J.B. Drake, "Classical and weak formulation of Stefan Problems with change of density upon change of phase", *Oak Ridge National Laboratory Report*, to appear.

[2] Bird, R.B., W.E. Stewart and E.N. Lightfoot, **Transport Phenomena**, Wiley, New York, 1960.

[3] Carlslaw, H.S. and J.C. Jaeger, **Conduction of Heat in Solids**, 2nd ed., Oxford U. Press, London, 1959.

[4] Drake, J.B. "Modeling convective Marangoni flows with void movement in the presence of solid-liquid phase change", *Oak Ridge National Laboratory Report, ORNL-6516* , Jan. 1990.

[5] Drake, J.B. "Convection in the melt", Ph.D. Thesis, Univ. of Tennessee, 1991.

[6] Wichner, R.P., A.D. Solomon, J.B. Drake and P.T. Williams, "Thermal analysis of heat storage canisters for a solar dynamic space power system", *Proceedings ASME Solar Energy Division Conference* , April 1988.

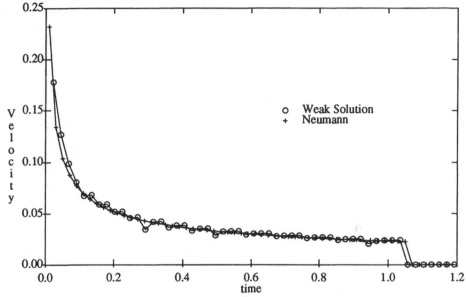

Figure 2: Calculated Velocity at x=0.4875

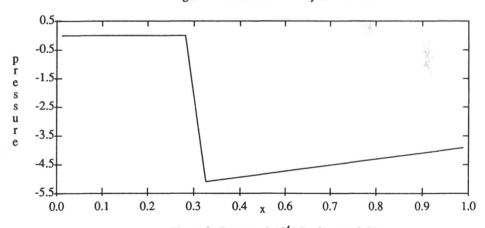

Figure 3: Pressure ($\times 10^4$) Profile at t=0.56

Vasilios Alexiades   and   *John B. Drake*

Mathematics Department

The University of Tennessee

Knoxville, TN 37996-1300

Mathematical Sciences Section

Oak Ridge National Laboratory

Oak Ridge, Tennessee 37831-6367

B V BAZALIY, I I DANILYUK AND S P DEGTYAREV

Classical solvability of some free boundary problems for parabolic equations with degeneration

We consider two problems unified by a common method of investigation which consists in a detailed study of corresponding nonclassical model boundary problems for linear partial differential equations.

I. In the filtration problem let two unmixed fluids at each moment $t \in [0, T]$ occupy regions $\Omega_t^{\pm} \subset R^n, n \geq 1$. Let $\Gamma_t = \bar{\Omega}_t^+ \cap \bar{\Omega}_t^-$, $\Sigma_T = \{(x, t) : x \in \Gamma_t, t \in [0, T]\}$. The surface Σ_T can be also represented as $\{(x, t) : x = x(\omega) + \bar{n}(\omega)\rho(\omega, t), t \in [0, t]\}$, where ω are some coordinates on $\Gamma_o, x(\omega) \in \Gamma_o$, $\bar{n}$ is a normal vector to Γ_o directed inside Ω_o^+, $\rho(\omega, t)$ is the function giving the deviation of Σ_T from the determined cylindrical surface $\Gamma_T = \Gamma_o * [0, T]$. One has to determine the functions $u^{\pm}(x, t)$ (pressure) and free surface Σ_t by some boundary conditions on a fixed surface, initial data $u^{\pm}(x, 0)$ and Γ_o, equations

$$u_t^+ - \Delta u^+ = f^+(u^+), \quad (x, t) \in Q_t^+ = \{(x, t) : x \in \Omega_t^+, t \in (0, T)\}$$

$$-\Delta u^- = f^-(u^-), \quad (x, t) \in Q_t^- = \{(x, t) : x \in \Omega_t^-, t \in (0, T)\}$$

and conditions on $\Sigma_T : u^+ = u^-$, $c_o \cos(\bar{N}, t) = \sum_{i=1}^n \cos(\bar{N}, x_i) u_{x_i}^{\pm}$. Thus the pressure distribution u^+ in Ω_T^+ is described by the parabolic equation and $u^-(x, t)$ in Q_T^- by the elliptic equation. Just in this case the degeneracy of the problem consists in.

Let us define the Banach space $\Pi^{i+\alpha}(\bar{\Omega}_T), i = 1, 2, E^{2+\alpha}(\bar{\Omega}_T)$ and $P^{2+a}(\bar{\Omega}_T)$ resulting from the closure of infinitely differentiable functions according to the norms:

$$|u|_{\Pi^{1+\alpha}(\bar{\Omega}_T)} \equiv |u|_{\Omega_T}^{(1+\alpha)} + [u]_{\Omega_T}^{(\alpha(1+\alpha)/2)} + [u]_{\Omega_T}^{(\alpha, \alpha/2)}$$

$$|u|_{\Pi^{2+\alpha}(\bar{\Omega}_T)} \equiv |u|_{\Omega_T}^{(2+\alpha)} + [u_x]_{\Omega_T}^{(\alpha(1+\alpha)/2)} + [u_{xx}]_{\Omega_T}^{(\alpha, \alpha/2)}$$

$$|u|_{E^{2+\alpha}(\bar{\Omega}_T)} \equiv |u|_{\Omega_T}^{(1+\alpha)} + [u_x]_{\Omega_T}^{(1+\alpha)} + [u_x]_{\Omega_T}^{(\alpha,(1+\alpha)/2)} + [u_{xx}]_{\Omega_T}^{(\alpha, \alpha/2)}$$

$$|u|_{P^{2+\alpha}(\bar{\Omega}_T)} \equiv |u|_{\Pi^{2+\alpha}(\bar{\Omega}_T)} + |u_t|_{\Pi^{2+\alpha}(\bar{\Omega}_T)}$$

Here $|\;|_{\Omega_T)}^{(1)}$ is a norm at $H^{1,1/2}(\bar{\Omega}_T)$ and the halfnorm

$$[u]_{\Omega_t}^{(\alpha,\beta)} = \sup_{x,y,t,\tau} |x-y|^{-\alpha}|t-\tau|^{-\beta}|u(x,t) - u(y,t) - u(x,\tau) - u(y,\tau)|$$

Spaces $\Pi^{i+\alpha}, E^{2+\alpha}, P^{2+\alpha}$ consist of elements of $\Pi^{i+\alpha}, E^{2+\alpha}, P^{2+\alpha}$ such that both the functions and their derivatives with respect to t become equal to zero when $t = 0$.

Theorem 1.

Let us assume that the initial data in the filtration problem is rather smooth, compatability conditions and $(\nabla u^+(x,0), \bar{n}) - (\nabla u^-(x,0), \bar{n}) > 0$ on Γ_0 are fulfilled. Then there is a $T_0 > 0$ depending on the problem data such that the solution

$$u^+(x,t) \in \Pi^{2+\alpha}(\bar{Q}_T^+), \ u^-(x,t) \in E^{2+\alpha}(\bar{Q}_T^-),$$

$$\rho(\omega,t) \in P^{2+\alpha}(\Gamma_T), 0 < T \leq T_0$$

exists.

While proving Theorem 1, it is of primary importance to study the following model conjugation problem. One has to find the functions $u^{\pm}, \rho$ given by

$$u_t^+ - a^+\Delta u^+ = f_1^+ f_{2z_i}^+, \ (z,t) \in R_{n,T}^+ = \{(z,t) : z_n > 0, t \in (0,T)\}$$

$$\epsilon u_t^- - a^-\Delta u^- = f_1^- f_{2z_i}^-, \ (z,t) \in R_{n,T}^- = \{(z,t) : z_n < 0, \ t \in (0,T)\} \qquad (1)$$

$$i = 1, n_i; \ \rho_t + k^{\pm}u_{z_n}^{\pm} + \sum_{i=1}^{n-1} b_i^{\pm}\rho_{z_i} = F_i^{\pm}(z,t), \ u^+ - u^- + d\rho = F_2(z,t)$$

$$z \in R_{n-1,T} = \{(z,t) : z_n = 0, \ t \in (0,T)\}$$

$$u^+ \in \Pi^{2+\alpha}(R_{n,T}), u^- \in E^{2+\alpha}(R_{n,T}^-), \ \rho \in P^{2+\alpha}(R_{n-1,T})$$

where $0 \leq \epsilon \leq \epsilon_0 = const, b^{\pm} \in R^{n-1}, a^{\pm}, k^{\pm}, d$ are positive constants.

Lemma 1. Problem (1) has the unique solution for which the following estimate is valid:

$$|u^+|_{\Pi^{2+\alpha}(R_{n,T}^+)} + |u^-|_{E^{2+\alpha}(R_{n,T}^-)}|\rho|_{P^{2+\alpha}(R_{n-1,T})} \leq$$

$$c\big(|f_2^+|_{\Pi^{1+\alpha}}R[f_1^+] + |f_2^-|_{\Pi^{1+\alpha}}R[f_1^-] + |F_1^+|_{\Pi^{1+\alpha}} + |F_1^-|_{\Pi^{1+\alpha}} + |F_2|_{P^{2+\alpha}}\big)$$

$$R[f] = <f>_t^{(1+\alpha)/2} + max|\nabla f| + max|f|$$

where c depends only on $a^{\pm}, k^{\pm}, b_i^{\pm}, \alpha, T$, sizes support right sides in (1) and does not depend on ϵ.

It follows from the boundary conditions at $z \in R_{n-1,T}$ that smoothness of ρ_t is equivalent to that of $u_t^{\pm}$, so the problem (1) is the conjugation problem with high order derivatives at boundary condititons and small parameter $\epsilon \leq O$ at high order derivative in one of the equations. For the proof of Lemma 1 we get the estimates of volume and surface potentials in the above classes uniform in ϵ, and then the investigation of the model problem is reduced to the study of a second kind Fredholm integral equation with zero kernel and cokernel. It should be noted that the essentially similar problem at $\epsilon > O$ arises in the study of the Stefan problem and the Stefan problem with impurity diffusion.

After passing to the problem in the fixed region and linearization, the initial problem is reduced to that of finding the fixed points of some nonlinear mapping. Using Lemma 1 we prove contractability of this mapping at sufficiently small time. which is the proof of Theorem 1.

II. Let us consider onedimensional two phase Stefan problem for quasilinear degenerating parabolic equations. In preset region $Q_T = \{(x,t) : x \in (-L_1, L_2), t \in (0,T)\}$ it is necessary to determine functions $u_i(x,t)$ and $\rho(t)$ by the conditions;

$$\phi_i(u_i)|u_i|^{n_i-1}u_{it} - (|u_i|^{m_i-1}u_{ix})_x = 0, \quad (x,t) \in Q_{i,T}, \ i = 1,2$$

$$u_i((-1)^i L_i, t) = f_i(t), \ u_i(x,0) = u_{0i}(x), \ \rho(0) = 0 \qquad (2)$$

$$u_i(\rho(t),t) = 0, \ \rho'(t) = |u_2|^{m_2-1}u_{2x} - |u_1|^{m_1-1}u_{1x}, \ x = \rho(t)$$

where $|\ln \phi_i(u_i)| \leq C$ for $u_i \in R^1$, $f_i(t)$, $u_{oi}(x)$ are the preset functions from the Holder classes, $f_1 \geq \epsilon > 0$, $f_2 \leq -\epsilon < 0$, $(-1)^{i+1}u_{0,i}(x) > 0$ for $x \neq 0$, $n_i \cdot m_i > 0$, $Q_{i,T} = Q_T \cap \{(x,t) : (-1)^i(x - \rho(t)) > 0, \ t \in (0,T)\}$.

Let us introduce spaces of functions in the region $\Omega_T = \{(x,t) : x \in \Omega, \ t \in (0,T)\}$

$$H_{\alpha}^{2+\beta,(2+\beta)/q}(\bar{\Omega}_T) = \{u(x,t) : \max_{\bar{\Omega}_T} |u| + <u>_{t,\Omega_T}^{(2+\beta)/q} +$$

$$|u_{xx}|_{\Omega_T}^{(\beta,\beta/q)} + |x^{\alpha}u_t|_{\Omega_T}^{(\beta,\beta/q)} < \infty\}, \ q = 2 + \alpha, \ \beta < \alpha$$

$$H_{-\alpha}^{q+\beta,(q+\beta)/q}(\bar{\Omega}_T) = \{u(x,t) : \max_{\bar{\Omega}_T} |u| + |u_t|_{\Omega_T}^{(\beta,\beta/q)} +$$

$$|x^{\alpha}u_{xx}|_{\Omega_T}^{(\beta,\beta/q)} < \infty\}, \ q = 2 - \alpha, \ 0 < \alpha < 1$$

90

Theorem 2

Problem (2) has a unique smooth solution at some interval $[0, T]$, here $\rho(t) \in C^{1+\gamma}([0, T])$, $|u_i|^{m_i-1} u_i \in H_\alpha^{2+\beta,(2+\beta)/q_i}(\bar{\Omega}_{i,T})$ at $\alpha_i = n_i/m_i - 1 \geq 0$ and $|u_i|^{m_i-1} u_i \in H_{-\alpha}^{q_i+\beta,(q_i+\beta)/q_i}(\bar{\Omega}_{i,T})$ at $\alpha_i = n_i/m_i - 1 < 0$, where $\gamma > 0$ is determined only by the values β, n_i, m_i, where $f_i(t), u_{0i}(x)$ are the traces of functions from the corresponding classes, compatibility conditions up to the first order included and $u_{0i_x}(0) \neq 0$ are fulfilled. At $m_i < n_i < 2m_i$ problem (2) has a unique solution at any interval $[0, T]$.

The local in time existence theorem is based on exact estimates of the boundary problem solution for the equation

$$x^\alpha u_t - u_{xx} + c(x,t)u_x + d(x,t)u = f(x,t)$$

in classes $H_\alpha^{2+\beta,(2+\beta)/q}(\bar{\Omega})$ and the equation

$$u_t - x^\alpha u_{xx} + c(x,t)u_x + d(x,t)u = f(x,t)$$

in classes $H_{-\alpha}^{q+\beta,(q+\beta)/q}(\bar{\Omega})$ where $\Omega_T = \{(x,t) : x \in (0,1),\ t \in (0,T)\}$ Global solvability in time is obtained using the proof of existence in the a priory estimate of the Holder constants for the first derivatives of the solution.

REFERENCES

[1] Bazaliy B.V., Danilyuk I.I., Degtyarev S.P. Classical solvability of the multidimensional time-dependent free boundary filtration problem. Dokl. Acad. USSR, scr.A.-1987.-N2.-p.3-7.(in Russian).

2.Bazaliy B.V., Degtyarev S.P. Solvability of problem with unknown boundary between domains of the parabolic and elliptic equations. Ukrain. math. journ.-1989. -v4i.N1O.-p.1343-1349 (in Russian).

3.Bazaliy B.V., Degtyarev S.P. Solvability of multidimensional free boundary filtration problem. Preprint 89-10. Donetsk, Inst. Appl Math. and Mech. Acad USSR. 1989. 49p. (in Russian).

4.Bazaliy B.V., Degtyarev S.P. The degenerating parabolic equations and free boundary problems. Dokl. Acad. USSR, ser.A.-1990.-N1,-p.3-7.(in Russian).

B. V. Bazaliy, I. I. Danilyuk and S. P. Degtyarev.
Institute of Applied Mathematics and Mechanics
of the Ukrainian Academy of Sciences.

G CAGINALP*

The issues of anisotropy and travelling wave velocity selection in the phase field model

1. The Basic Equations The relationship between the phase field equations and the sharp interface problems which constitute its distinguished limits has been explored in detail [C1]. If we consider the limit of the surface tension and kinetics model (modified Stefan) then these equations have the form, with T as (absolute) temperature and ϕ as the order parameter,

$$\rho c_p T_t + \rho \frac{\ell}{2} \phi_t = K \triangle T \tag{1}$$

$$\alpha \epsilon^2 \phi_t = \epsilon^2 \triangle \phi + \frac{1}{2}(\phi - \phi^3) + \frac{\epsilon}{3} \frac{[s]_E}{\sigma}(T - T_M) \tag{2}$$

in a region $\Omega \subseteq \mathbb{R}^d$. Here $\rho :=$ density, $c_p :=$ specific heat, $\ell :=$ latent heat, $K :=$ thermal conductivity, $\sigma :=$ surface tension, $[s]_E :=$ entropy density difference between phases, $T_M =$ (absolute) melting temperature and ϵ is the width of the interface. Thus, ϵ is physically on the order of Angstroms $(= 10^{-8} cm)$ and the mathematical limits of sharp interface problems are attained as $\epsilon \to 0$. The numerical studies, however, are carried out by "stretching out" the width of the interface, so that $\epsilon >> 10^{-8} cm$, with the ansatz that this spreading does not appreciably alter the development of the interface [CS 1, 2]. (Note that other scalings have also been used in numerics (e.g. [FL], [K]).

If ϵ approaches zero with all other parameters held fixed, then equations (1), (2) are governed to leading (macroscopic) order by the (sharp interface) surface tension and kinetics model,

$$\rho c_p T_t = K \triangle T \quad \text{in} \quad \Omega \backslash \Gamma \tag{3}$$

$$\rho \ell v = K \nabla T \cdot n]_+^- \quad \text{on} \quad \Gamma \tag{4}$$

*Supported by NSF Grant DMS-9002242 .

$$[s]_E(T - T_M) = -\rho\kappa - \alpha\sigma v \quad \text{on } \Gamma \tag{5}$$

where Γ is now the (sharp interface which arises from the level set of $\phi = 0$ in (1), (2).

2. The Issue of Anisotropy In order to consider anisotropic properties of materials, the equations (1), (2) need to be rederived or modified in one of at least three ways.

(A) The original free energy $\mathcal{F}$ which led to (2) can be rederived from microscopic considerations [C2]. This is accomplished by writing the lattice Hamiltonian term $\sum_{x,x'} J(x - x')\Phi(x)\Phi(x')$, where J represents the interaction strength between N "spins" Φ. One can prove the identity

$$N \sum_{x}, x' J(x - x')\Phi(x)\Phi(x') = \sum_{q} \hat{J}(q)\hat{\Phi}(q)\hat{\Phi}(-q) \tag{6}$$

where $\hat{J}(q) := \sum_{x} e^{-iq \cdot x} J(x)$, $\hat{\Phi}(q) := \sum_{x} e^{iq \cdot x}\Phi(x)$ in terms of the wave number q. For small q, one can expand $\hat{J}(q) \simeq c_0 + c_2 q^2 + c_4 q^4 + \cdots$, where the odd terms vanish for lattices of appropriate symmetry. The term $\sum_{q} q^2 \hat{\Phi}(q)\hat{\Phi}(-q)$ leads to the $(\nabla\phi)^2$ in $\mathcal{F}$ which then forms $\epsilon^2 \triangle\phi$.

The anisotropic effects of J will be reflected in the coefficients c_2, c_4, etc. However, each of theses coefficients picks up some of the anisotropy and averages the remainder. Thus, the c_2 terms will manifest 2-fold or 3-fold anisotropy but will average out more detailed anisotropy. The examination of detailed anisotropy thus entails the study of higher order equations [CF].

(B) Another possibility is to modify the $(\nabla\phi)^2$ term in the free energy directly, without an explicit derivation from microscopic physics. This modifies (2) into

$$\alpha\epsilon^2\phi_t = \epsilon^2 g(\theta)\triangle\phi + \frac{1}{2}(\phi - \phi^3) + \frac{\epsilon}{3}\frac{[s]_E}{\sigma}(T - T_M) \tag{7}$$

where θ is the angle between the normal to the interface, i.e. $\nabla\phi$, and a fixed angle θ_o which is the preferred direction. For 2-fold or 3-fold anisotropy this will be identical to (A).

For more detailed anisotropy, one can choose a form of $g(\theta)$ which will lead to the correct macroscopic equation in a way similar to that of (C) below.

(C) The most phenomenological alternative is the direct modification of σ and α in (2) into

$$\alpha(\theta)\epsilon^2\phi_t = \epsilon^2\triangle\phi + \frac{1}{2}(\phi - \phi^3) + \frac{\epsilon}{3}\frac{[s]_E}{\sigma(\theta) + \sigma''(\theta)}(T - T_M). \tag{8}$$

The system (1), (8) has been studied numerically [CS2] and exhibits the growth of physically realistic crystals.

The asymptotic limit of (1), (8) is (3), (4) and

$$[s]_E(T - T_M) = -[\sigma(\theta) + \sigma''(\theta)](\kappa + \alpha(\theta)v) \tag{5'}$$

We note an interesting feature all three approaches in that anisotropy of "equilibrium properties" such as the surface tension, σ, influences the "dynamical anisotropy" namely the coefficient of v in (5') which is often denoted $\beta^{-1}(\theta)$ and called mobility. It was first demonstrated directly from the microscopics in [C2] that assuming anisotropic interactions would not only lead to an anisotropic term multiplying the curvature (the Gibbs-Thomson condition) but would also yield a related term for the mobility. Since the dynamics arises from the evolution equation $\alpha \phi_t = -\dfrac{\delta \mathcal{F}}{\delta \phi}$, one might have guessed that an isotropic α would lead to an isotropic mobility. For the computation of practical problems in which experimental data suggests the form of $\beta(\theta)$, one can readily define $\alpha(\theta)$ so that $[\sigma(\theta) + \sigma''(\theta)]\alpha(\theta) = \beta(\theta)$.

3. **Travelling Waves for Phase Field and Sharp Interfaces** The scaling of the phase field equations is particularly important for travelling waves. The advantage of (1), (2) over other possible scalings is that it disentangles the limit $\sigma \to 0$ from $\epsilon \to 0$. This allows a direct comparison between the diffused interface model and the sharp interface model when all other macroscopic parameters (particularly surface tension) are identical. The result [CN] can be summarized briefly by stating that for sufficiently small ϵ, there is no difference between (1), (2) and (3) - (5) in terms of the existence of travelling waves.

When $\alpha \neq 0, \sigma \neq 0$ and the undercooling (at x = ∞) is large, then (3) - (5) has a solution of the form $u(x - v^*t)$ and $\Gamma(t) = v^*$, where $v^* =$ constant is determined by the undercooling (e.g. [CC]). Under these conditions, (1), (2) also has a travelling wave solution [CN] if ϵ is sufficiently small. Its velocity $v(\epsilon)$ approaches v^* as $\epsilon \to 0$ [CN]. If $\sigma = 0$ then neither (3) - (5) nor (1), (2) (for small ϵ) have such solutions.

Hence, one may conclude that the $\epsilon \to 0$ limit is <u>not</u> a singular perturbation unlike the surface tension limit $\sigma \to 0$ which is very singular, even within this one-dimensional setting. As noted in [CC] it is this mobility term [$\alpha\sigma \neq 0$ in(5)] which <u>selects</u> the velocity as a function of undercooling. Note that the regularity of the ϵ-perturbation is a prerequisite for meaningful numerics, since practical computational constraints and macroscopic length scales of 1 to$10^{-1}cm$ imply that the value of ϵ in the numerical calculations will be much larger than the physical value of $\epsilon \sim 10^{-8}cm$.

An alternative scaling of the phase field equations entails a proportionality relation between the surface tension and interface thickness and appeared in some of the earlier papers on the subject ([C3], [F]). Since the capillarity length, $d_o :=$ $c_p \sigma / (\ell[s]_E)$ is also very small ($\sim 10^{-6} cm$) this is a reasonable scaling which does not have [(3) - (5)] as a distinguished limit.

Recently, several works have considered this scaling in the context of travelling waves and other fronts [LBT], [BS].

REFERENCES

[BS] M. Barber and D. Singleton, *Travelling waves in phase field models of solidification*, in preparation.

[C1] G. Caginalp, *Mathematical models of phase boundaries*, Material Instabilities in Continuum Problems and Related Mathematical Problems (J. Ball, ed.), Symposium 1985-1986, Heriot-Watt, Oxford, 1988, pp. 35-52.

[C2] _____, *The role of microscopic anisotropy in the macroscopic behavior of a phase boundary*, Annals of Physics **172** (1986), 136-155.

[C3] _____, *The limiting behavior of a free boundary in the phase field model*, Carnegie-Mellon Research Report 82-5 (1982).

[CC] J. Chadam and G. Caginalp, *Stability of interfaces with velocity connection term*, [Univ. of Pittsburgh Preprint 1987], Rocky Mountain Journal of Mathematics **21** (1991), 617-629.

[CF] G. Caginalp and P. Fife, *Higher order phase field models and detailed anisotropy*, Physical Review B **34** (1986), 4940-4943.

[CN] G. Caginalp and Y. Nishiura, *The existence of travelling waves for phase field equations and convergence to sharp interface models in the singular limit*, Quarterly of Applied Math **49** (1991), 147-162.

[CS1] G. Caginalp and E. Socolovsky, *Efficient computation of a sharp interface by spreading via phase field methods*, Applied Math. Letters **2** (1989), 117-120.

[CS2] _____, *Efficient computation of interfaces and instabilities by smoothing via phase field methods*, Advances in Mathematics, Computation and Reactor Physics, 1991.

[F] G. Fix, in Free Boundary Problems (A. Fasano and M. Primicerio, eds.), Pitman, London, 1983.

[FL] J. Lin and G. Fix, in Theory Methods Appl. **12** (1988), 811.

[K] R. Kobayashi, *Private communication*.

[LBT] H. Lowen, J. Bechhoefer and L. Tuckerman, *Crystal growth at long times: critical behavior at the crossover from diffusion to kinetics limited regimes*, Simon Fraser University Preprint.

G CAGINALP AND J JONES
Phase field equations with fluid properties

1. INTRODUCTION

The phase field equations in their simplest form are a parabolic system of equations which describe the temperature and an order or phase parameter. This approach has provided an alterntive to the sharp interface type models in which the interface has explicit conditions imposed upon it. In fact the phase field system provides a means of deriving these conditions from the underlying physics [1, 2].

In this paper we extend the phase field model to more complicated physical situations in which fluid properties are incorporated within the context of a unified and consistent derivation. Large systems of equations which are brought together from physically different analyses often suffer from internal contradictions which limits their usefulness. In our derivation of the equations below, we have attempted to make the derivation from basic free energies as uniform as possible within the constraints of maintaining a manageable system of equations. The details of this derivation are being reported in [8]. One of the main conclusions of our analysis is the generalization of the classic Gibbs-Thomson relation to one which relates the temperature at the interface to a set of fluid and thermal variables and coefficients.

2. THE EQUATIONS

We let $T(t, x)$ be the difference between the temperature and the equilibrium temperature at a time $t \in \mathbb{R}^+$ and $x \in \Omega \subset \mathbb{R}^d$. Similarly $P(t, x)$ is the pressure difference. In place of the usual phase or order paramenter $\varphi(t, x)$, we define $\lambda(t, x)$ which has an interpretation as the mass fraction of liquid when the system is in equilibrium so that λ near 0 is solid while λ near 1 is liquid. The other variables are are $w(t, x)$, the specific volume (= 1/density), $\vec{u}(t, x)$, the fluid velocity, and $S(t, x)$,

Supported by NSF Grant DMS-9002242 and NIST Grant 60NANBOD1026 .

96

the entropy density. We use the subscript E to denote values at the phase coexistence curve in the $P - T$ plane and let $[w]_E \equiv \{w \text{ (solid equil)} - w \text{ (liquid equil)}\}$, etc. We define the physical constants

$$\bar{\epsilon}\eta \equiv \text{coefficient of thermal expansion} = \frac{\partial w}{\partial T}\Big|_{P,\lambda}$$

$$\bar{\epsilon}^2 \nu \equiv \text{isothermal compressibility} = \frac{-\partial w}{\partial T}\Big|_{T,\lambda}$$

$T_E, \ P_E \equiv$ absolute temperature and pressure at coexistence

$$C_E \equiv \text{coexistence heat capacity} = T_E \frac{\partial S}{\partial T}\Big|_{w,\lambda}$$

$D, K \equiv$ phase and thermal diffusivities

$$\tilde{\ell} \equiv \left\{ \frac{\eta}{\nu} \left[[\eta]_E - \frac{\eta}{\nu}[\nu]_E \right] T + [S]_E \right\}(T + T_E)$$

$=$ effective latent heat

$A \equiv$ rate constant

$\bar{\epsilon} \equiv$ scaling parameter which approaches zero

The effective latent heat $\tilde{\ell}$ accounts for the work performed in compressing (or expanding) the material, in addition to the heat evolved in the phase transition.

We use the standard fluid mechanics notation $\dfrac{d}{dt} \equiv \dfrac{\partial}{\partial t} + \vec{u} \cdot \vec{\nabla}$ for the total time derivative (which measures the time change of a "particle" of fluid). The heat diffusion equation is then

$$C_E \frac{dT}{dt} + \bar{\epsilon}^{-1}\frac{\eta}{\nu}(T + T_E)\, w\vec{\nabla} \cdot \vec{u} + \tilde{\ell}\frac{d\lambda}{dt} = K\triangle T \tag{1}$$

while the equation of continuity and Newton's law are given by

$$\frac{dw}{dt} = w\vec{\nabla} \cdot \vec{u} \text{ and } \frac{d\vec{u}}{dt} = -w\vec{\nabla}P. \tag{2), (3}$$

The equation of state is in the form of an equation for w,

$$w = (T, P, \lambda) = w_E(\lambda) + \bar{\epsilon}\eta(\lambda)T - \bar{\epsilon}^2\, \nu(\lambda)P, \tag{4}$$

which includes the possibility of compression and expansion due to thermal and mechanical effects. This linear equation may be interpreted as the leading terms in a series expansion about the coexistence curve of the exact equation of state.

The change in phase is governed by the equation

$$\frac{d\lambda}{dt} = \bar{\epsilon}\, D\, \triangle\lambda + \bar{\epsilon}^{-1}\, A\, \alpha\, \lambda(1 - \lambda)(\lambda - \frac{1}{2})$$

$$+ A^{-1}\left\{ \frac{\eta}{\nu}\left[[\eta]_E - \frac{1}{2}\frac{\eta}{\nu}[\nu]_E \right] T^2 + [S]_E\, T \right\} \tag{5}$$

The phase diffusion constant D is $O(\bar{\epsilon}^{-1})$ so that $D \equiv d\bar{\epsilon}^{-1}$ where $d = O(1)$. It is convenient to define a new parameter ϵ by

$$\epsilon^2 \equiv \bar{\epsilon}^2 \alpha^{-1} A^{-1} D = \bar{\epsilon} \alpha^{-1} A^{-1} d \tag{6}$$

so that the phase equation (5) can be rewritten as

$$\frac{\epsilon^2}{d} \frac{d\lambda}{dt} = \epsilon^2 \triangle \lambda + \alpha \lambda (1 - \lambda)(\lambda - \frac{1}{2}) + \frac{\epsilon}{d^{1/2}} \left\{ \frac{\eta}{\nu} \left[[\eta]_E - \frac{1}{2} \frac{\eta}{\nu} [\nu]_E \right] T^2 + [S]_E T \right\} \tag{7}$$

Note that the term $\alpha \lambda (1 - \lambda)(\lambda - \frac{1}{2})$ is a prototype of a derivative of a double well potential which depends on microscopic properties. This term can be replaced by a more general function $f(\lambda)$ which has the same qualitative features while incorporating the detailed microscopic structure of the transition region.

Thus, equations (1) - (5) are a complete system $T, \vec{u}, w, \lambda, P$ and can be studied subject to appropriate initial and boundary conditions.

3. BASIC ASYMPTOTICS AND NEW INTERFACE RELATION

A key question which arises within this formulation involves the relationship between temperature and the other variables. The latter now includes not only surface tension and curvature (as in the Gibbs-Thomson relation [3]) and velocity (as in models which incorporate kinetic undercooling [3, 4]) but pressure, density, velocity of fluid in addition to physical constants such as the coefficients of thermal expansion, and isothermal compressibility, etc.

An asymptotic analysis for a plane wave case results in the new interfacial relation

$$A_1 T(0)^2 + A_2 T(0) + \frac{1}{2d^{1/2}} [v - u(0)]\sigma = 0 \tag{8}$$

where σ is the surface tension, v the interface velocity and

$$A_1 \equiv \frac{\eta}{\nu} \left[[\eta]_E - \frac{1}{2} \frac{\eta}{\nu} [\nu]_E \right], \qquad A_2 \equiv [S]_E.$$

Thus, the additional nonlinearities introduced in this system of equations result in a nonlinear equation (in temperature) which replaces the linear equations in temperature and velocity. For nonplanar interfaces, the relation (8) also includes a term, $\kappa \sigma$, where κ is the sum of principal curvatures.

For the plane wave problem, one can obtain a set of relations in which the other variables are written in terms of the temperature and integration constants C_1, C_2 which are determined by boundary conditions through a matching procedure and are not necessarily of order 1. In particular, one has

$$u = -C_1(A_1 T^2 + A_2 T) + C_2 \tag{9}$$

$$w = -A_1 T^2 - A_2 T + C_1^{-1} C_2 - C_1^{-1} v \tag{10}$$

$$P = -C_1\{-C_1 A_1 T^2 - C_1 A_2 T + C_2\} + v \tag{11}$$

To leading order, the phase variable is given by

$$\lambda = \frac{1}{2}\{\tanh{(z/2\epsilon)} + 1\} \tag{12}$$

where $z \equiv x - vt$.

The study of phase boundary problems with fluid properties has also been undertaken within the sharp interface context [5 - 7].

REFERENCES

1. G. Caginalp , *An analysis of a phase field model of a free boundary*, Archive for Rational Mechanics and Analysis **92** (1986), 205-245; see also *Limiting behavior of a moving boundary in the phase field model* Carnegie-Mellon Research Report 82-5 (1982).
2. _____, *The role of microscopic anisotropy in the macroscopic behavior of a phase boundary*, Annals of Physics **172** (1986), 136-155.
3. B. Chalmers , *Principles of Solidification*, R.E. Krieger Publ., Huntington, NY (1977).
4. G. Horvay and J.W. Cahn , Acta Met. 9 (1961), 695.
5. V. Alexiades and J.B. Drake , *A weak formulation for phase change problems with bulk movement due to unequal densities*, in proceedings of this conference.
6. H.W. Alt , See contribution to this conference.
7. I. Pawlow , See contribution to this conference.
8. G. Caginalp and J. Jones , *A phase-fluid model: derivation and new interface relation*, in IMA Volumes on Phase Transitions (1990).

G. CAGINALP AND J. JONES

Mathematics Department
University of Pittsburgh
Pittsburgh, PA 15260 USA

G CAGINALP AND E A SOCOLOVSKY

A unified computational approach to phase boundaries by spreading: single needle, crystal growth and motion by mean curvature

ABSTRACT. The phase field approach can be used to approximate a wide range of sharp interface problems including the classical Stefan model, the surface tension model, the Cahn-Allen model, motion by mean curvature, etc. The parameters of a single set of equations can be adjusted so that one can vary continuously from one model to another. The sharp interface is smoothed out within a scaling which preserves the most important physical quantities. The two-dimensional and radially symmetric n-dimensional computations indicate that this efficient method for dealing with these stiff problems results in very accurate interface determination without interface tracking. We use these methods to (i) provide a numerical verification of the concept of a critical radius and quantitative agreement; (ii) qualitative agreement with single needle crystals, anisotropic crystal growth and motion by mean curvature; (iii) verification of constant velocity dendritic growth (independent of initial conditions).

I. Introduction. Numerical computations of moving boundaries have posed important and difficult problems (see [1, 9] for surveys). For such problems, a variety of techniques are used. These include interface tracking, regularization, "smoothing" of the interface, and numerous methods designed for special purposes.

In this paper we present a computational technique for a broad class of free boundary problems based on the ideas of the phase field approach. We apply this technique to a set of problems which arise from phase transitions and accurately determine the interface without tracking it separately. In particular, Stefan-type models, with or without surface tension, and other effects are approximated very accurately and efficiently with a smooth system of parabolic equations (see [2] and

Supported by NSF Grant DMS-9002242 and NIST GRANT 60NANBOD1026 .

references in [3]). These ideas are the numerical counterpart of the theory intro-duced in Section 4 of [4].

A feature relevant to efficiency is that the width of the interface (and consequently the stiffness of the equations) can be changed. This results in execution times which are reduced by more than one order of magnitude without significant change in the evolution of the interface. This conclusion has been corroborated qualitatively and by comparison with an exact solution. In fact, one can obtain an accuracy of four digits with an interface diffused to one-fifth of the entire domain. Also see [10].

We have also applied these concepts to study a well-known instability in materi-als science, namely the unstable equilibrium at critical undercooling, and to model crystal growth under kinetic anisotropy (see Section IV). Our results provide a nu-merical verification of the onset of this instability and confirm the critical nature of the magnitude of the surface tension and provide for the approximation of mo-tion by mean curvature. Even in this subtle situation, the interface thickness can be modified without significantly altering the results. Other aspects of numerics involving the phase field model have been investigated in [5-7, 11].

II. The sharp interface models and the phase field model. The computa-tion of a sharp interface problem using phase field methods can be illustrated on the surface tension and kinetics model descried below. We seek a function $T(x, t)$, rep-resenting the absolute temperature, and a curve $\Gamma(t) \subset \Omega$, representing the interface such that

$$\rho c_p T_t = K \Delta T \qquad \text{in } \Omega \backslash \Gamma(t) \tag{2.1}$$

$$\rho \ell v = K \nabla T \cdot \hat{n}]_+^- \qquad \text{on } \Gamma(t) \tag{2.2}$$

$$[s]_E(T - T_M) = -\sigma \kappa - \alpha \, \sigma \, v \qquad \text{on } \Gamma(t) \tag{2.3}$$

where $\rho :=$ density, $c_p :=$ specific heat, $\ell :=$ latent heat per mass, $K :=$ thermal conduc-tivity, $[s]_E :=$ entropy density difference between phases, $T_M :=$ (absolute) melting temperature, $\sigma :=$ surface tension, $\alpha :=$ relaxation constant (i.e. $[s]_E/(\alpha\sigma) :=$ mobil-ity). The sum of principal curvatures of $\Gamma(t)$ at a point x is denoted κ while $\hat{n}$ is the unit normal, v the velocity and $]_+^-$ is the jump from solid to liquid. The classical Stefan model is attained from (2.1) - (2.3) by setting σ to zero.

The phase field equations have the form

$$\rho c_p T_t + \rho \frac{\ell}{2} \phi_t = K \Delta T \tag{2.4}$$

101

$$\alpha \epsilon^2 \phi_t = \epsilon^2 \Delta \phi + \frac{1}{2}(\phi - \phi^3) + \frac{\epsilon}{3} \frac{[s]_E}{\sigma}(T - T_M) \tag{2.5}$$

where ϕ is the phase or order parameter. Thus the coefficients above have all been identified except for ϵ which is related to the physical width of the interface. In the distinguished limit as ϵ approaches zero while all other parameters are held fixed, the solutions to (2.4), (2.5) are governed by those of (2.1) - (2.3) to leading order. The initial and boundary conditions for ϕ are determined by the conditions on u. For example, Dirichlet boundary condtions for ϕ must be chosen so that $\phi = \phi_\pm$ on Ω where $\phi_\pm$ are the left and right roots of

$$\frac{1}{2}(\phi - \phi^3) + \frac{\epsilon}{3} \frac{[s]_E}{\sigma}(T - T_M) = 0 \tag{2.6}$$

where $T(x)$ is the temperature on the boundary.

III. Spreading the sharp interface. The smoothing of any of the problems of the form [(2.1) - (2.3)] is accomplished by fixing the physical parameters σ and α unless either one is zero. Then one adjusts ϵ, thereby changing the interfacial thickness and the "stiffness" of (2.5), while holding σ fixed in (2.3). One must ensure that the algebraic equation (2.6) still has three distinct roots, but beyond this constraint, one is free to choose ϵ. Of course, the smaller the ϵ, the closer the approximation to [(2.1) - (2.3)] and also the more points that are needed in order to compensate for the stiffness of the problem. If h denotes the mesh size, empirically we found that a ratio $\epsilon/h \approx 1$ provides optimal accuracy.

We consider a two dimensional or a spherically symmetric problem in an annular geometry. Results for the planar front modified Stefan problem, with different values for ϵ, are displayed in Figure 1. With the computer and package used for these computations, an interface 36 times wider reduced the CPU time by a factor of 560. Notice that the stiffness of equations (2.4) - (2.5) involves a term proportional to ϵ^{-2}. We then observe that the interfacial thickness can be changed considerably without a significant difference in the development of the interface. Numerical trials also show that even a small change in the surface tension, σ, affects the motion of the interface.

The consistency observed above can be confirmed by comparing our numerical results with the exact solution for a planar classical Stefan front. In this case the approximation is possible by taking small values for both σ and ϵ. For a typical melting or freezing problem (see Figure 2) we find an agreement of three to four digits even when the ϵ is such that the interface transition takes about 20% of the domain. For a transition of about 35% of the domain, the accuracy is close to two digits.

figure1
Modified Stefan Problem
surface tension (sigma): 0.085333
epsilons (top to bottom) are: .04608, .02048, .00128

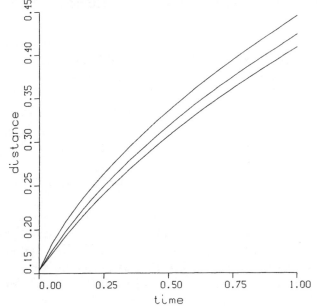

Figure 2
Exact and computed Stefan interface
Grid size: .002 - Max. Relative Error: 0.31%
sigma= .00533 - epsilon= .00128 - a= .16 - psi= .0032

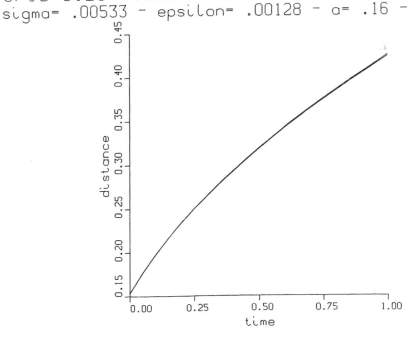

Note that in a planar front problem the curvature is automatically zero. However, the surface tension, σ, still plays an important role in (2.3) because of the kinetic undercooling term $-\alpha\sigma v$. The computations differ significantly as σ is changed and α is kept constant.

IV. Critical radius, crystal growth and motion by mean curvature. At the interface, the velocity, curvature and temperature are related by

$$v = -\frac{1}{\alpha}(\kappa + \frac{[s]_E}{\sigma}(T - T_M)) \qquad (4.1)$$

It follows that if a solid sphere of curvature $\kappa_0 \equiv (d-1)/r_0$ (d is the space dimension and r_0 its radius) is surrounded by its melt and the surface tension is σ_0, then equilibrium will prevail when there is a constant temperature of T_o such that $\kappa_0 = [s]_E(T_M - T_0)/\sigma$. This is an unstable equilibrium configuration which is well known (see [8] p.67) to melt or freeze upon varying any of the parameters. We have obtained numerical confirmation of the onset of this instability and observe it by varying κ_0 or σ_0 slightly. For instance, taking an initial radius r larger than the critical radius r_0, we have that $v > 0$, i.e. freezing occurs. However, a much larger change in the interfacial thickness does not alter the direction or (approximate) magnitude of the interface velocity. This set of numerics confirms the theoretical and experimental analysis of materials science. Details are discussed in [10].

In a different set of experiments, portrayed in Figures 3 to 6, we used an identical intitial interface with four sets of parameters representing very different physical problems. The intitial interface shape is given in radial coordinates by $r = 0.3(1 + (cos8\theta)/3)$. In Figure 3, the parameters are adjusted so that the value of the surface tension, σ_0, corresponds to a critical radius of $r_0 \equiv 0.433$. As indicated by (4.2) the velocity is then positive and freezing occurs with an interface that evolves toward a spherical shape.

Next, we implement the parameters which approximate the Cahn-Allen motion by mean curvature. In this limit, the role of temperature is trivialized by letting ℓ approach zero so that only (2.5) applies with $T - T_M = 0$. The limiting equation is then $v = -\kappa/\alpha$. The results, shown in Figure 4, indicate that the interface moves in response to curvature, so that the intitial shape becomes spherical and shrinks. In Figure 5, dendritic growth is obtained using α which is anisotropic concentrated in a band around the diagonal. Finally, in Figure 6 we show the formation of a crystal when α is locally anisotropic with the diagonals as crystallographic directions. In our experiments the final evolution shapes were very stable since they were also obtained under many other intitial conditions and physical parameters. The computations and graphics metafiles for Figures 3 to 6 were obtained at the Pittsburgh Supercomputer Center.

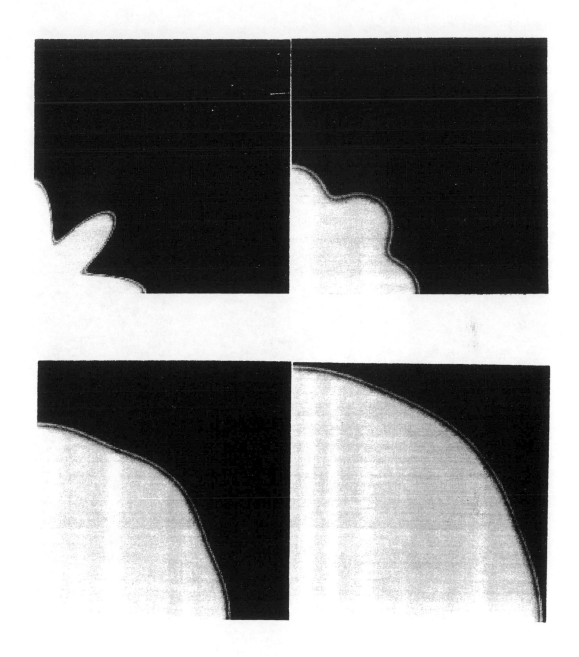

FIGURE 3

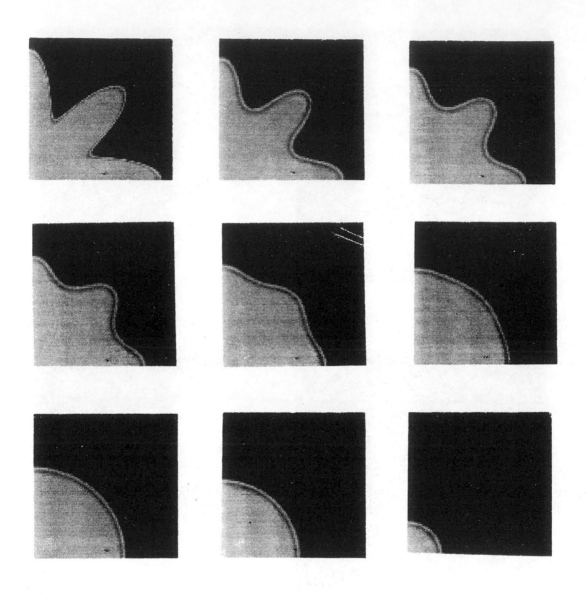

FIGURE 4

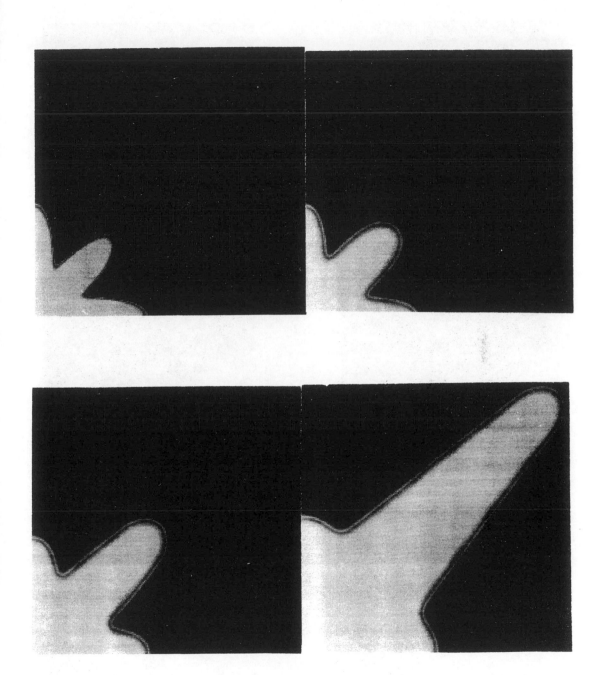

FIGURE 5

107

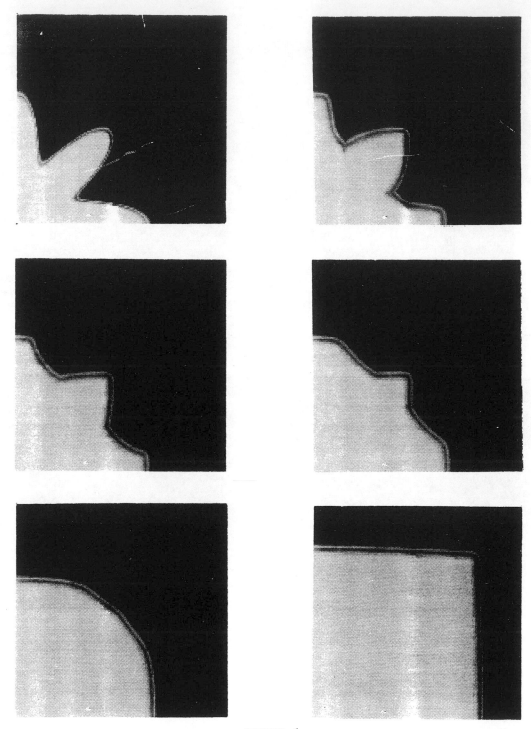

FIGURE 6

It is well known from experimental and theoretical studies that the velocity of the dendrite is constant after an initial acceleration. This constant velocity is independent of the initial interfacial shape. Other numerical experiments confirmed these features with a high degree of accuracy.

REFERENCES

1. J. Glimm, *The continuous structure of discontinuities*, Courant Inst. Preprint.
2. G. Caginalp, *Surface tension and supercooling in solidification theory*, Lecture Notes in Physics, vol. 216, Springer, Berlin, 1984, pp. 216-226.
3. _____, *Stefan and Hele-Shaw type models as asymptotic limits of the phase field equations*, Physical Review A **39** (1989), 5887-5896.
4. _____, *Mathematical models of phase boundaries*, Symposium on Material Instabilities in Continuum Mechanics (J. Ball, ed.), Heriot-Watt Univ. (1985-1986), Oxford Science Publ., 1988, pp. 35-50.
5. J. Lin and G. Fix, *Numerical simulations of nonlinear phase transitions I. The isotropic case*, Nonlinear Analysis, Theory, Meth. Appl **12** (1988), 811-823.
6. G. Caginalp and J. Lin, *A numerical analysis of an anisotropic phase field model*, IMA Journal of Applied Math **39** (1987), 51-66.
7. A. Visintin, *Surface tension effects in phase transitions*, in [4].
8. B. Chalmers, *Principles of Solidification*, Krieger Publ., New York.
9. R.M. Furzeland, *A comparative study of numerical methods for moving boundary problems*, J. Inst. Math. Appl. **126** (1980), 411-429.
10. G. Caginalp and E.A. Socolovsky, *Computation of sharp pahse boundaries by spreading: The planar and spherically symmetric cases*, Journal of Computational Physics **95** (1991), 85-100.
11. M. Mimura, R. Kobayashi and H. Okazaki, *Computer simulations of instabilities and dendritic growth using phase field model*, Videotape.

G. CAGINALP AND E.A. SOCOLOVSKY

Mathematics Department
University of Pittsburgh
Pittsburgh, PA 15260

G CAGINALP AND W XIE

The transition between unbounded and bounded velocities in sharp interface models

INTRODUCTION The purpose of this paper is to discuss the transition between blow-up of solutions in the classical Stefan model and finite velocities in the surface tension-kinetics model. The phase field model can be used to understand the continuous transition between these two regimes. Asymptotic analysis suggests that as surface tension, σ, becomes smaller, the velocity increases (under conditions to be specified below) until $v(T) \to \infty$ for $\sigma \to 0$. Thus, our analysis suggests that the crucial singular perturbation in blow-up involves the limit $\sigma \to 0$.

The phase field equations are a parabolic system describing the (dimensionless) temperature, $u(x,t)$, and an "order parameter" or phase field, $\varphi(x,t)$ at point x and time $t > 0$. The phase field, φ, is scaled so that φ near $+1$ is associated with the liquid phase and φ near -1 with the solid. The interface is defined implicitly as the set of points for which φ vanishes. Further discussion can be found in [1] and references therein.

The basic phase field equations can be written as

$$\alpha\xi^2\varphi_t = \xi^2\Delta\varphi + a^{-1}g(\varphi) + 2u \tag{1.1}$$

$$u_t + \frac{\ell}{2}\varphi_t = K\Delta u \tag{1.2}$$

where ℓ and K are positive constants representing latent heat and diffusivity, respectively. The function g is a derivative of a symmetric double well potential with minima at ± 1, e.g. $g(\varphi) = \frac{1}{2}(\varphi - \varphi^3)$, and ϵ and a are small parameters. The equation (1.1), (1.2) can be studied subject to initial conditions

$$u(x,0) = u_0(x), \varphi(x,0) = \varphi_0(x) \quad x \in \Omega \tag{1.3}$$

* Supported by NIST Grant 60NANBOD1026 and NSF Grant DMS-9002242 .

and appropriate boundary conditions such as

$$\frac{\partial u}{\partial n} = 0, \ \varphi(x,t) = \varphi_\pm \quad x \in \partial\Omega \tag{1.4}$$

where $\varphi_\pm$ are the right and left roots of $a^{-1}g(\varphi) + 2u = 0$ (these roots will be approximately ± 1 since a is small).

As we can see from [4], the parameter, defined by

$$\epsilon = \xi a^{1/2}, \ \sigma = \xi a^{-1/2} \tag{1.5}$$

are related to the length scale which measures the width of the transition layer between solid and liquid and to the surface tension, respectively. The sharp interface problems can be approached by phase field equations if the parameters ϵ and/or σ approach to zero in the sense of asymptotic limits.

We will discuss the possible blow-up case for the solution of phase field equations as the parameters ϵ and σ approach to zero. This is essentially related to the fact that the solution of classical Stefan problem will blow-up for certain initial data.

In the next section we will present a sufficient condition such that the blow-up will occur for the solutions of one-dimensional classical Stefan problem.

2. BLOW-UP FOR THE CLASSICAL STEFAN PROBLEM

We consider the solution $(s(t), u(x,t))$ of the one dimensional Stefan problem such that

$$u_t = Ku_{xx} = \begin{cases} k_\ell u_{xx} & \text{in } Q_1 = \{(x,t)|0 < x < s(t), t > 0\} \\ k_s u_{xx} & \text{in } Q_2 = \{(x,t)|s(t) < x < 1, t > 0\} \end{cases} \tag{2.1}$$

$$u_x(i,t) = 0 \quad i = 0,1 \tag{2.2}$$

$$u(x,0) = u_0(x) = \begin{cases} \psi_1(x), & 0 \le x \le b \\ \psi_2(x), & b \le x \le 1, \end{cases} \tag{2.3}$$

and with the interface condition of $x = s(t)$ $(s(0) = b)$

$$u = 0 \tag{2.4}$$

$$k_\ell u_x^- - k_x u_x^+ = -L\dot{s}, \tag{2.5}$$

where $0 < b < 1$ is a constant. In the case of $\psi_2 = 0$, it is a one phase problem and the necessary and sufficient condition for the finite time blow-up can be found in [3], which is entirely determined by the quantity $\int_o^b (\psi_1(x) + L)dx < 0$. If, in

111

addition, $\psi_1(x)$ satisfies $-u_0 \leq \psi_1 \leq 0$. with $\psi_1'' \geq 0$. $\psi_1'(b) \geq \dfrac{u_o}{b}$ and $u_0 > L > o$, then we have that $\dot{s}(t) \to -\infty$ as $t \to T^-$, where

$$T = \frac{L(\int_0^b (u_0 + \psi_1)dx)^2}{ku_0^2(u_0 - L)}.$$

In fact. if we consider the quantity $Q(t) = \displaystyle\int_0^{s(t)} (u + u_0)dx$, then one can deduce that $Q \geq \dfrac{ku_0^2}{-2L\dot{s}}$ and $\dfrac{dQ^2}{dt} \leq \dfrac{ku_0^2(u_0 - L)}{-L}$. Hence

$$Q^2(t) \leq (\int_0^b (u_0 + \psi_1)dx)^2 + \frac{ku_0^2(u_0 - L)}{-L}t.$$

This implies that $t \leq T$ and $-\dot{s}(t) \geq \dfrac{ku_0^2}{2Q(t)} \to \infty$ as $t \nearrow T^-$.

In the two-phase case, we can choose the initial data in the several ways such that the finite time blow-up occurs.

If we choose the initial data such that $\psi_i \leq 0$ $(i = 1.2)$, (i.e. the liquid is supercooled), in this case the free boundary is monotone decreasing. Note that for the solution $(s; u)$, we have

$$Ls(t) = Q - \int_o^1 u(x,t)\ dx \tag{2.6}$$

$$Q = L\,b + \int_o^b \psi_1 dx + \int_b^1 \psi_2 dx. \tag{2.7}$$

Then we can see that, if the solution exist globally. it is $Q \geq 0$ (since finite extinction implies $Q = \displaystyle\int_0^1 u(x,t)dx \leq 0$, otherwise $\displaystyle\lim_{t\to\infty} s(t) = Q > 0$). Thus, if we choose ψ_1, ψ_2 such that $(M_1(x - b) \leq \psi_1(x) \leq 0, \quad -M_2 x \leq \psi_2(x) \leq 0$, say, with $Q < \dfrac{-M_2}{2} < 0$, then the finite time blow-up will occur since neither the global existence nor finite time extinction can occur in this case. The particular choice we can make is that $\psi_1 = M_1(x - b)$ and $\psi_2 = -M_2(x - b)$ with $M_2(2 - b) + 2L < M_1 b$ ($M_1 = 6L, M_2 = \dfrac{L}{3}$ and $b = \dfrac{1}{3}$ say).

If we choose the initial data such that $\psi_1 \leq 0$ and $\psi_2 \geq 0$ (i.e. the liquid is supercooled and solid is superheated). we can choose ψ_1, ψ_2 $(M_1(x - 1) \leq \psi_1 \leq 0, \quad 0 \leq \psi_2 \leq M_2 x$, say) such that $Q < L - \dfrac{M_1}{2} < 0$. Then for the reason demonstrated above, the finite time blow-up will occur in this case. The particular choice we can make is $\psi_1 = M_1(x-1)$ if $0 \le x \le \frac{1}{4}$, $\psi_1 = 3M_2(x - \frac{1}{2})$ if $\frac{1}{4} \le x \le \frac{1}{2}$ and $\psi_2 = \frac{M_2}{4}(x - \frac{1}{2})$ with $M_1 > 2L$ and $M_2 + 6M_1 < 16L(M_1 = \frac{5}{2}L, M_2 = \frac{L}{2}$, say).

It is clear that the other possible cases can be discussed in a similar way.

3. APPROXIMATING SHARP INTERFACE PROBLEMS

As shown in [1], the formal asymptotic limits of the phase field equations depending on each scaling regime are given as follows: it approaches the classical Stefan model (resp. surface tension kinetics model) in the limit of $\epsilon/\sigma, \sigma \to 0$ (resp. $\epsilon \to 0$, σ held fixed) for a fixed positive α. We now consider the phase field equation (1.1) - (1.4) with $u_0(x)$ as the particular case chosen in Section 2 and suitable $\varphi_0(x)$ with $\varphi_\pm = \pm 1$. Note that, under these initial and boundary conditions there exists a unique global solution to (1.1) - (1.3) [4]. The interface, $\Gamma(t)$, defined by

$$\Gamma(t) = \{x | \varphi(x,t) = 0\} \tag{3.1}$$

is regular provided the initial and boundary conditions are smooth.

Let $u(x,t,\sigma,\epsilon), \varphi(x,t,\sigma,\epsilon)$ be functions with internal transition layer with width $0(\epsilon)$ at $x = s(t,\sigma,\epsilon)$. As in [1, 2], we set up outer and inner expansions for the function u, φ and s. Thus

$$u = u(x,t,\sigma,\epsilon) = u^0(x,t,\sigma) + \epsilon u^1(x,t,\sigma) + \cdots$$

with similar expressions for φ and s.

The inner expansion proceeds by defining $z = \dfrac{x - s}{\epsilon}$ and thinking of u and φ as depending in a regular manner on the variables z and t near $\Gamma(t)$:

$$u = U(z,t,\sigma,\epsilon) = U^0(z,t,\sigma) + \epsilon U^1(z,t,\sigma) + \cdots$$

with a similar expression for φ.

It was shown in [1] tht, as $\epsilon \to 0$, (s^0, u^0) satisfies

$$u_t^0 = ku_{xx}^0 \text{ in } Q_1 \cup Q_2, \tag{3.2}$$

$$u^0 = -0(\sigma)\alpha s_t^0 \text{ on } x = s^0, \tag{3.3}$$

$$k((u_x^0)^- - (u_x^0)^+) = -\ell s_t^0 \text{ on } x = s^0, \tag{3.4}$$

$$u_x^0(i,t) = 0 \quad i = 0, 1, \tag{3.5}$$

113

$$u^0(x,0) = \psi(x) \ \ 0 \leq x \leq 1, \tag{3.6}$$

where $\psi(x)$ was chosen in Section 2 and $s^0(0) = b$.

We note that, for the problem (3.2) - (3.6), there exists a unique solution such that $s^0(t,\sigma) \in C^\infty(0,T]$ [5]. We now claim that, for the solution (s^0, u^0), there exists $t^*(0 < t^* < T)$ such that $|s^0_t(t,\sigma)| \to \infty$ as $\sigma \to o^+$ and $t \nearrow t^*$. Indeed, if this is not true, then by a calculation of the integrals $\displaystyle\int_o^T \frac{d}{dt} \int_0^1 (u^0)^2 \ dx dt$ and $\displaystyle\int_0^T \frac{d}{dt} \int_0^1 K(u^0_x)^2 \ dx dt$ and using boundary interface conditions, we have

$$\sup_{0 \leq t \leq T} \int_0^1 [(u^0(x,t,\sigma))^2 + (u^0_x(x,t,\sigma))^2]dx+$$

$$\iint_{Q_1^\sigma \cup Q_2^\sigma} [(u^0_x)^2 + (u^0_{xx})^2] \ dx dt \leq C[\|\psi\|^2_{H^1(0,1)}]$$

where $Q_1^\sigma = \{(x,t)|0 < x < s^0(t,\sigma), 0 < t < T\}, Q_2^\sigma = \{(x,t)|s^0(t,\sigma) < x < 1, 0 < t < T\}$ and C is independent of σ.

By the compactness argument, we see that there exists a couple $(s^0(t), u^0(x,t))$ with $s^0(t) \in C^{0,1}_{[0,1]}$ and $u \in C^{\alpha,\frac{\alpha}{2}}(\overline{Q}_T)(0 < \alpha < 1), (Q_T = \{0 < x < 1, \sigma < t < T\})$ such that $s^0(t,\sigma) \to s^0(t)$ uniformly in $C[0,T]$ and $u^0(x,t,\sigma) \to u^0(x,t)$ uniformly in $C(\overline{Q}_T)$ as $\sigma \to 0$; moreover, $(s^0(t), u^0(x,t))$ satisfies (3.2) (3.5), (3.6) and

$$u^0 = 0 \text{ on } x = s^0(t) \tag{3.7}$$

$$\ell s^0(t) = \ell b + \int_0^1 \psi(x) \ dx - \int_0^1 u^0(x,t) \ dx. \tag{3.8}$$

Note that $s^0(t) \in C^{0,1}[0,T]$ and then $u^0_x \in C(\overline{Q}_1) \cap C(\overline{Q}_2)$, then (3.8) implies the Stefan condition (3.4). Thus $(s^0(t); u^0(x,t))$ is a classical solution of (2.1) - (2.5), but this clearly contradicts the fact that for this particular initial data $\psi(x)$, the soluition of (2.1) - (2.5) will blow-up.

The above demonstration suggests that, with the parameters $\xi, a \to 0$ (with $\sigma \to 0$), the velocity of the internal transition layer of the phase field model will get larger and larger for some initial data of u, while the velocity remains bounded for any choice of initial data $u_0(x)$ as the parameter $\xi, a \to 0$ (with σ and α held fixed). This is actually true for s^0, the zero order of the expansion. One needs to investigate the behavior of s^1 to confirm this. Many technical details will be involved in this case, and we conjecture that the result remains true for s^1.

4. THE GLOBAL PICTURE - TRANSITION FROM FINITE VELOCITY TO BLOW-UP AS A FUNCTION OF PARAMETERS

The results discussed in the previous sections indicate that the phase field equations (for fixed ϵ and σ) result in bounded velocities while the limiting Stefan problem ($\epsilon \to 0, \sigma \to 0$) results in unbounded velocities at finite time T (under suitable initial and boundary conditions). This situation leads to the questions of the scaling of crossover behavior between the two regimes. Sketching the velocity against time as shown in Figure 1, one has a divergence at $t = T$ for the classical Stefan model. The analogous curves for the phase field model do not diverge but should attain higher values at T as σ and ϵ approach zero, as illustrated. Hence, one can try to find a function $h(\sigma, \epsilon)$ which governs the crossover from the finite crossing of $t = T$ to the divergence at $t = T$. The structure of h may involve a power law dependence on critical exponents e.g. $h \sim \sigma^p \epsilon^q$.

A slightly different view is represented by Figure 2 in which the velocity at $t = T$ is plotted against σ and ϵ resulting in a surface in three-dimensional space. Here, we know that there is a divergence at $\sigma = \epsilon = 0$ and the asymptotics indicate that there should be a surface of the form displayed in Figure 2. In terms of this picture, the level curves of constant velocity at $t = T$ would provide an alternative view of the scaling relations.

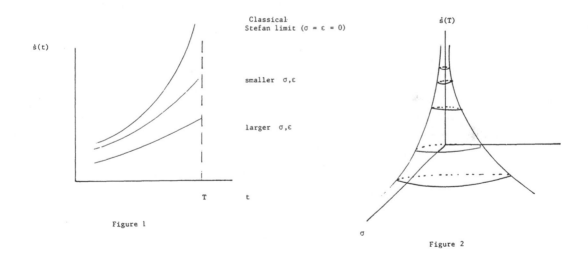

Figure 1

Figure 2

115

References

[1] G. Caginalp , *Stefan and Hele-Shaw typemodels as asymptotic limits of the phase field equations*, Physical Review A **39** (1988), 35-50, Symposium on Material Instabilities in Continuum Mechanics (J. Ball, ed.), Heriot-Watt Univ. (1985-1986), Oxford Science Publ.; see also Mathematical models for phase boundaries.

[2] G. Caginalp and P.C. Fife , *Dynamics of layered interface arising from phase boundaries*. SIAM J. Appl. Math **48** (1988), 506-518.

[3] A. Fasano, M. Primicerio , *New results on some classical parabolic free-boundary problems*, Quart. Appl. Math. **38** (1981), 439-460.

[4] G. Caginalp , *An analysis of a phase field model of a free boundary*, Arch Rat. Mech. Anal. **92** (1986), 205-245.

[5] W. Xie , *The Stefan problem with a kinetic condition at the free boundary*, SIAM J. Math. Anal. **21** (1990), 362-373.

G. CAGINALP AND W. XIE

Mathematics Department
University of Pittsburgh
Pittsburgh, PA 15260 USA

J M CHADAM AND H-M YIN

The two-phase Stefan problem in materials with memory

Abstract   In this paper we consider a phase transition in materials with memory. A suitable model is formulated for such a process as a free boundary problem associated with a Volterra integrodifferential equation. By considering a suitable approximate problem and using integral estimates we prove the existence and uniqueness of the weak solution to the problem. Moreover, the regularity of the weak solution is also investigated.

1. Introduction It is well-known (cf. [3] and [8]) that an ice-water melting process can be described by the enthalpy equation:

$$\frac{\partial H(u)}{\partial t} = \triangle u$$

where

$$H(u) = \begin{cases} a_1 u, & \text{if } u > 0, \\ [-1, 0], & \text{if } u = 0, \\ a_2 u - 1, & \text{if } u < 0. \end{cases}$$

is a monotone graph, while the temperature of the phase change is assumed to be 0 and the specific heat is normlized to be 1. However, in some materials with memory, Fourier's law is apparently no longer suitable. One needs to take into account the effect of the past history. Then the relationship between the temperature and the heat flux takes the following form (cf. [6] and [7]):

$$\vec{q}(x, t) = -a \nabla u(x, t) - \int_0^{+\infty} b(\tau) \nabla u(x, t - \tau) d\tau. \tag{.2}$$

If one further assumes that the internal energy depends linearly on the temperature, conservation of energy leads to the following integrodifferential equation:

$$u_t = div[\, a \nabla u] + \int_0^{+\infty} b(\tau) \triangle u(x, t - \tau) d\tau$$

which can be written as

$$u_t = a\Delta u + \int_0^t b(t-\tau)\Delta u(x,\tau)d\tau + f_0(x,t),$$

where

$$f_0(x,t) = \int_{-\infty}^0 b(t-\tau)\Delta u(x,\tau)d\tau$$

is assumed to be known. Without loss of generality, we may assume that $f_0(x,t) = 0$ in the sequel.

When a phase transition takes place in such a system, unlike the classical Stefan problem, the solid and the liquid phases are actually determined by the sign of $u(x,t) + \int_0^t b(t-\tau)u d\tau$ while the interface (free boundary) is the graph of the set

$$\{(x,t) \in \bar{Q}_T : u(x,t) + \int_0^t b(t-\tau)u(x,\tau)d\tau = 0\}.$$

This leads by the balance of energy to the following enthalpy formulation:

$$\frac{\partial H(u,t)}{\partial t} = \Delta u + \int_0^t b(t-\tau)\Delta u(x,\tau)d\tau, \qquad (1.1)$$

where $H(u,t)$ is defined by

$$H(u,t) = \begin{cases} a_1 u, & \text{if } u + \int_0^t b(t-\tau)u d\tau > 0, \\ [-1,0], & \text{if } u + \int_0^t b(t-\tau)u d\tau = 0, \\ a_2 u - 1, & \text{if } u + \int_0^t b(t-\tau)u d\tau < 0. \end{cases} \qquad (1.2)$$

We shall point out that $H(u,t)$ may depend on the time varible since in our case the phase change takes place whenever $u(x,t) + \int_0^t b(t-\tau)u d\tau = 0$ rather than that of $u(x,t)$. Our objective in this paper is to study the above equation (1.1) in $Q_T = \{(x,t) \in R^2 : 0 < x < 1, 0 < t < T\}$ subject to the initial and boundary conditions:

$$
\begin{aligned}
u(0,t) &= f_1(t), & 0 \leq t \leq T, & \qquad (1.3) \\
u(1,t) &= f_2(t), & 0 \leq t \leq T, & \qquad (1.4) \\
u(x,0) &= u_0(x), & 0 \leq x \leq 1. & \qquad (1.5)
\end{aligned}
$$

For simplicity, we call the system (1.1)-(1.5) as P(A).
To define a solution of P(A), we let

$$B = \{\psi \in C^{2,1}(\bar{Q}_T) : \psi(0,t) = \psi(1,t) = \psi(x,T) = 0\}$$

be a test Banach space. A weak solution of P(A) can now be defined as follows:

Definition : A continuous function $u(x,t)$ defined on $\bar{Q}_T$ is called a weak solution of P(A), if $u(x,t)$ satisfies the following integral identity:

$$\int_0^T \int_0^1 \{H(u,t)\psi_t + [u + \int_0^t b(t-\tau)u\,d\tau]\psi_{xx}\}dx\,dt$$

$$= \int_0^T [f_2(t)\psi_x(1,t) - f_1(t)\psi_x(0,t)]dt - \int_0^1 u_0(x)\psi(x,0)dx \qquad (1.6)$$

for any test function $\psi(x,t) \in B$, where $H(u,t)$ is defined by (1.2).

The study of various Stefan-like problems has received much attention in the past decades. Recently, to avoid infinite speed of propagation of heat, some authors have considered the Stefan problem for hyperbolic equations (cf. [4], [5], [9], etc.). These models are, however, unable to take the into consideration the effect of past history. Our formulation seems to be suitable for the phase change of heat flow in materials with memory. The propagation speed of heat is, however, still infinite.

We list some hypotheses which we will use in following sections.

H(1): The functions $f_1(t), f_2(t) \in C^2[0,T]$, $u_0(x) \in C^{2+\alpha}[0,1]$ and $b(t) \in C^{1+\alpha}[0,T]$;

H(2): $f_1(0) = u_0(0), f_2(0) = u_0(1), a_1 f_1'(0) = u_0''(0), a_2 f_2'(0) = u_0''(1)$.

To obtain a classical solution, we need certain further sign restrictions on the known data:

H(3): $f_1(t) \geq c_0 > 0$ and $f_2(t) \leq -c_0$. There exists a $s_0 \in [0,1]$ such that $u_0(s_0) = 0, u_0(x)(x - s_0) < 0$ for $x \in [0, s_0) \bigcup (s_0, 1]$ and $u_0'(s_0) < 0$.

The existence and uniqueness of the weak solution is proved in Section 2. The regularity is studied in Section 3.

2. The Weak Solution

To prove the existence of a weak solution, we consider the following approximate problem:

$$\frac{\partial H_n(u,t)}{\partial t} = u_{xx} + \int_0^t b(t-\tau)u_{xx}d\tau, \quad in\ Q_T, \qquad (2.1)$$

$$u(0,t) = f_1(t), \qquad\qquad 0 \leq t \leq T, \qquad (2.2)$$

$$u(1,t) = f_2(t), \qquad\qquad 0 \leq t \leq T, \qquad (2.3)$$

$$u(x,0) = u_0(x), \qquad\qquad 0 \leq x \leq 1, \qquad (2.4)$$

where $H_n(\xi, t) \in C^3(R^1 \times [0,T])$ and satisfies for any $t \in [0,T]$:

(i) $\quad H_{n\xi}(\xi, t) > \dfrac{1}{2}min\{a_1, a_2\} > 0, H_{nt}(\xi, t)$ is uniformly bounded;

(ii) $\quad H_n(\xi, t) = H(\xi, t)$ if $\xi \leq 0$ or $\xi \geq \dfrac{1}{n}, n = 1, 2, \cdots$;

(iii) $\quad H_n(\xi, t) \to H(\xi, t)$ in $L^2(-A_0, A_0)$ for any $A_0 > 0$ as $n \to +\infty$.

By the results of [10], there exists a unique solution $u_n(x,t) \in C^{2+\alpha, 1+\frac{\alpha}{2}}(\bar{Q}_T)$. We now derive several uniform estimates for $u_n(x,t)$. Observe that the classical maximum principle is invalid for our equation (2.1). By the integration by parts and application of Gronwall's inequality, we can deduce the following estimate .

Lemma 2.1: There exists a constant C_1 such that

$$\int_0^T \int_0^1 u_{nt}^2 \, dx \, dt + \sup_{0 \leq t \leq T} \int_0^1 u_{nx}^2(x, t) dx \leq C_1, \qquad (2.5)$$

where C_1 depends only on the known data.

As a direct consequence, we have

Corollary 2.1: There exists a constant C_2 which depends only on the known data such that if $(x_i, t_i) \in \bar{Q}_T$, $i = 1, 2$,

$$|u_n(x_1, t_1) - u_n x_2, t_2)| \leq C_3[|x_1 - x_2|^{\frac{1}{2}} + |t_1 - t_2|^{\frac{1}{4}}].$$

Once one has the above estimate, one can obtain the following result by a compactness argument which is similar to the classical Stefan problem (cf. [3] and [8]):

Theorem 2.1: There exists a unique weak solution for the problem (A) under the hypotheses H(1)-H(2).

3. The Regularity In this section we prove that the free boundary is composed of a Lipchitz curve. Throughout this section the conditions H(1)-H(3) are assumed. The method is similar to those of [1], [2] and [10]. It can be seen that the maximum principle for parabolic equations plays an essential role in the proofs of the theorems in [1] and [10]. This principle is, however, not true for

120

the solution u_n. To overcome this difficulty, we shall introduce a new function $U(x,t)$ defined by

$$U(x,t) = u_n(x,t) + \int_0^t b(t-\tau)u_n(x,\tau)d\tau, \qquad (x,t) \in \bar{Q}_T.$$

By using the Neumann series, we can express $u_n(x,t)$ in terms of $U(x,t)$ in following way

$$u_n(x,t) = U(x,t) + \int_0^t B(t,\tau)U(x,\tau)d\tau,$$

where $B(t,\tau)$ is a smooth known function depending only on $b(t-\tau)$ and T. Moreover, $U(x,t)$ satisfies

$$U_t + B(t,t)U + \int_0^t B_t(t,\tau)U\,d\tau \quad = \quad \frac{1}{H_{nu}(u_n,t)}\Delta U \; in \; Q_T \qquad (3.1)$$

$$U(0,t) \quad = \quad f_1(t) + b\int_0^t f_1(\tau)d\tau, \qquad 0 \le t \le T \quad (3.2)$$

$$U(1,t) \quad = \quad f_2(t) + b\int_0^t f_2(\tau)d\tau, \qquad 0 \le t \le T \quad (3.3)$$

$$U(x,0) \quad = \quad u_0(x), \qquad 0 \le x \le 1. \qquad (3.3)$$

It is clear that we can apply the maximum principle for the function $U(x,t)$. We shall analyze the structure of level sets of $U(x,t)$ rather than those of $u_n(x,t)$. Let ε be a noncritical value of $U(x,t)$. Moreover, we restrict that ε is small enough so that $0 < \varepsilon < \frac{1}{2}\min_{0\le t\le T} f_1(t)$. Hence, the level set $\{(x,t) \in \bar{Q}_T : U = \varepsilon\}$ is not empty. Sard's Lemma and the implicit function theorem along with the strong maximum principle implies the following results.

Lemma 3.1:

$$\{(x,t) \in \bar{Q}_T : U(x,t) = \varepsilon\} \quad = \quad \{(s_\varepsilon(t),t)\}$$
$$\{(x,t) \in \bar{Q}_T : U(x,t) = -\varepsilon\} \quad = \quad \{(s_{-\varepsilon}(t),t)\}$$

where the curves $s_\varepsilon(t)$ and $s_{-\varepsilon}(t)$ are differentiable and monotonic increasing in the time direction. Moreover, the two curves start at the bottom $t = 0$ and exit at the top $t = T$.

Furthermore, there exists a constant $E > 0$ independent of n but which may depend on ε such that

$$\frac{\partial U_x(s_\varepsilon(t),t)}{\partial x} < -E, \qquad \frac{\partial U_x(s_{-\varepsilon}(t),t)}{\partial x} < -E.$$

The proof of this lemma is similar to [1] since the strong maximum principle is valid for $U(x,t)$. Employing the above result, we can show

Theorem 3.1: There exists a Lipschitz continuous function $s(t)$ such that

$$
\begin{aligned}
\Gamma_T &= \{(x,t) : u(x,t) + \int_0^t b(t-\tau)u(x,\tau)d\tau = 0\} \\
&\equiv \{(x,t) : x = s(t),\ 0 \le t \le T\}. \text{ Moreover,} \\
s(t) &= s_0 + \int_0^t [-u_x^-(s(\tau),\tau) + u_x^+(s(\tau),\tau)]d\tau \\
&\quad + \int_0^t \int_0^\tau b(\tau-\xi)[-u_x^-(s(\xi),\xi) + u_x^+(s(\xi),\xi)]d\xi d\tau.
\end{aligned}
$$

Furhtermore,

$$
Q_T^L = \{(x,t) : U(x,t) > 0\} = \{(x,t) : 0 < x < s(t),\ 0 < t < T\},
$$

$$
Q_T^S = \{(x,t) : U(x,t) < 0\} = \{(x,t) : s(t) < x < 1,\ 0 < t < T\}.
$$

With the above results in hand we can obtain

Theorem 3.2: there exists a unique classical solution to P(A) under the conditions H(1)-H(2).

References

1. J. R. Cannon and H. M. Yin, On the existence of the weak solution and the regularity of the free boundary for a degenerate Stefan problem. J. Diff. Eqs., 73(1988), 104-118.

2. J. R. Cannon and H. M. Yin, A periodic free boundary problem arising in some diffusion chemical reaction processes, Nonlinear Analysis, TMA, to appear.

3. A. Friedman, The Stefan problem in several space variables, Trans. Amer. Math. Soc., 133(1968), 51-87.

4. A. Friedman and B. Hu, The Stefan problem for a hyperbolic heat equation, J. Math. Anal. Appl., 138(1989), 249-279.

5. J. M. Greenberg, A hyperbolic heat transfer problem with phase change, IMA J. Appl. Math., 38(1987), 1-21.

6. S.-O. Londen and J. A. Nohel, Nonlinear Volterra integrodifferential equations ocurring in heat flow, J. Integral Eqs., 6(1984), 11-50.

7. J. W. Nunziato, On heat conduction in materials with memory, Quart. Appl. Math., 29(1971), 187-204.

8. O. A. Oleinik, A method of solution of the general Stefan problem, Dokl. A.N. SSSR., 135(1960), 1054-1057.

9. R. R. Showalter and N. J. Walkington, A hyperbolic Stefan problem, Quart. Appl. Math., 45(1987), 769-781.

10. H. M. Yin, The Classical solutions for nonlinear integrodifferential equations, J. Integral Eq. and Appl., 1(1988), 249-263.

11. H. M. Yin, The classical solution of the periodic Stefan problem, J. Part. Diff. Eqs., 1(1988), Series A, 43-60.

John M. Chadam and Hong-Ming Yin
Dept. of Math. and Stat., McMaster Univ., Hamilton, Ontario,
L8S 4K1, Canada

J I DIAZ[(1)(2)], A FASANO[(*)] AND A MEIRMANOV

On the disappearance of the mushy region in multidimensional Stefan problems

1. Introduction. We present in this communication some of the results of the work by the authors [2], [3] concerning the disappearance in a finite time of the <u>mushy region</u> for the Stefan problem. Previous results in this direction was obtained in [8] for one-dimensional problems (see also [1]). For the treatment of the multidimensional problems, different cases must be introduced according the "nature" of the spatial domain Ω and the type of boundary conditions. Our results deals with this complex situation but rest on the same general program: we first consider the problem under symmetry conditions on the domain and the data and then we reduce the treatment of general formulations to the symmetry case by means of rearrangements techniques.

To fix ideas and to simplify the exposition we shall restrict ourselves to the consideration of the following one-phase problem (we send the reader to [2], [3] for the treatment of two-phases problem and other boundary conditions):

$$(1) \qquad \frac{\partial u}{\partial t} - \Delta\theta = 0 \qquad\qquad \text{in } Q_\infty = \Omega \times (0, \infty)$$

$$(2) \qquad \theta(x, t) = \theta^0(t) \qquad\qquad \text{on } \Sigma_\infty = \partial\Omega \times (0, \infty)$$

$$(3) \qquad u(x, 0) = u_0(x) \qquad\qquad \text{on } \Omega,$$

where Ω is a regular bounded open set of $\mathbb{R}^N$. As usual, u and θ represents the rescaled <u>entalhpy</u> and <u>temperature</u>, respectively, and they are related by

$$(4) \qquad u = \phi(\theta)$$

where

[(1)] Presenter.
[(2)] Partially Supported by the DGICYT project. PB86/0485.
[(*)] Partially supported by the Italian MURST and by the MMMI Project of the CNR

(5) $\begin{cases} \phi \text{ is a strictly increasing continuous function on } (0,\infty) \text{ extended} \\ \text{to } [0,\infty) \text{ by } \phi(0)=[-L,0]. \end{cases}$

Here L>0 represents the <u>latent</u> <u>heat</u>. We assume on the data the following conditions:

(6) $\quad u_0 \in L^\infty(\Omega), \qquad u_0(x) \geq -L \qquad \text{a.e. } x \in \Omega$

(7) $\quad \theta^0 \in W^{1,\infty}(0,\infty), \qquad \theta^0(t) \geq 0 \qquad \text{for any } t \geq 0 .$

Our results will be concerned with <u>weak</u> <u>solutions</u> i.e. pairs of functions $(u, \theta) \in [L^2_{loc}(0,\infty:L^2(\Omega)) \cap C([0,\infty):L^1(\Omega))] \times L^2_{loc}(0,\infty:H^1(\Omega))$ satisfying (2),(4) and such that

(8) $\quad \int_\Omega u(x,t)\eta(x,t)dx - \int_0^t \int_\Omega u(x,\tau)\eta_t(x,\tau)dxd\tau + \int_0^t \int_\Omega \nabla\theta \cdot \nabla\eta dxd\tau = \int_\Omega u_0(x)\eta(x,0)dx$

for any t>0 and any $\eta \in H^1(\Omega \times (0,t)), \eta(.,s) \in H^1_0(\Omega)$ if $0 \leq s \leq t$. Existence uniqueness and many other results on weak solutions are today well-known in the literature (see e.g. [5],[7],[10]). In particular we know that $u, \theta \in L^\infty(Q_\infty)$ and that $u(x,t) \geq -L$, $\theta(x,t) \geq 0$ a.e. (x,t) in Q_∞. So, for any $t \geq 0$, we have the domain decomposition $\Omega = l(t) \cup m(t)$ where $l(t)$ and $m(t)$ are the <u>liquid</u> <u>phase</u> and the <u>mushy</u> <u>region</u> defined by

$$l(t)=\{x \in \Omega: \theta(x,t)>0\} \qquad \text{and} \qquad m(t)=\{x \in \Omega: \theta(x,t)=0\}.$$

Our main goal is to obtain conditions on $\theta^0(t)$ and $u_0(x)$ in order to assure the disappearance (in measure) of region $m(t)$ after a finite time t^*.

2. The radially symmetric case. The treatment of the problem becomes easier under symmetry assumptions. So let $\Omega = B(0,r_0) = \{x \in \mathbb{R}^N: |x| < r_0\}$ and assume u_0 as before and such that

(9) $u_0(x)=u_0(|x|)$ satisfying $\begin{cases} u_0(r) = -L & \text{if } 0 \leq r \leq R_0 \\ u_0(r) > -L & \text{if } R_0 \leq r \leq r_0 \end{cases}$

for some $R_0 \in (0,r_0)$. In order to state our result we need to introduce some auxiliary notation. Given N>1 we define

(10) $\quad \varphi_N(r) = \begin{cases} r^{N-1}(N-2)^{-1}(r^{2-N}-r_0^{2-N}) & \text{if } N>2 \\ r\log(r/r_0) & \text{if } N=2 . \end{cases}$

Finally, given θ^0 and u_0 we define the quantities

(11) $$D(t:u_0, \theta^0) = \int_0^{r_0} u_0(r)\varphi_N(r)dr + \int_0^t \theta^0(\tau)d\tau$$

and

(12) $$D_\infty \equiv D(\infty:u_0, \theta^0) = \lim_{t\to\infty} D(t:u_0, \theta^0).$$

We have

Theorem. 1. *Let u_0 satisfying (9). Then*

(i)   If $D_\infty \leq 0$ the mushy region exists ($|m(t)|>0$) for any time $t\geq 0$.

(ii) If $D_\infty > 0$ there exists a finite time $t^*>0$ such that the mushy region disappear ($|m(t)|=0$) for any time $t\geq t^*$.

(iii) Assume $D_\infty > \gamma$ where

(13) $$\gamma = M\int_0^{r_0}\varphi_N(r)dr \quad , \quad M = \|u_0\|_{L^\infty(\Omega)} + \phi(\sup_{0<t}\theta^0(t)).$$

Then the estimate $t^*>t_0^*$ holds with $t_0^* \geq 0$ defined by

(14) $$D(t_0^*:u_0, \theta^0) = \gamma . \quad \blacksquare$$

3. The general case. To state the result for the general case of Ω an open bounded set of $\mathbb{R}^N$ we start by recalling the notion of increasing rearrangement of a function: Let $h \in L^1(\Omega)$; the <u>distribution</u> <u>function</u> of h is defined by $\mu(\tau) = |\{x\in\Omega: h(x)<\tau\}|$. The <u>increasing</u> <u>rearrangement</u> of h is the one-variable function $\tilde{h}: [0, |\Omega|] \to \mathbb{R}$ given by $\tilde{h}(s) = \inf\{\tau\in\mathbb{R}: \mu(\tau)>s\}$. Finally, the <u>increasing</u> <u>symmetric</u> <u>rearrangement</u> of h is the function $h^*: \Omega^* \to \mathbb{R}$ defined by

(15) $$h^*(x) = \tilde{h}(\omega_N|x|^N)$$

where Ω^* is the ball of $\mathbb{R}^N$, centered at the origin and with the same measure as Ω, i.e.

(16) $$\Omega^* = B(0, r_0) \text{ with } r_0 \text{ such that } |\Omega| = |\Omega^*| = \omega_N r_0^N,$$

where ω_N denotes the measure of the unit ball in $\mathbb{R}^N$.

Our main result concerning the mushy region is the following:

Theorem. 2. *Let Ω be a regular bounded open set of $\mathbb{R}^N$. Let u_0 and θ^0 satisfy (6) and (7) respectively. We also assume that*

(17) $$\theta^0(t) \qquad \text{is nondecreasing in t}$$

and

126

(18) $u_0(x) \leq \theta^0(0)$ a.e. $x \in \Omega$.

Finally, let $r_0 > 0$ be given by (16) and define $u_0(r)$ in $L^\infty(0, r_0)$ by

(19) $u_0(r) := \tilde{u}_0(\omega_N r^N)$.

Then the following conclusions hold:

a) *Let (U, Θ) be the solution of the Stefan problem on $Q_\infty^* \equiv \Omega^* \times (0, \infty)$ corresponding to initial value u_0^* and boundary data θ^0. Then*

(20) $|m(t)| \leq |M(t)|$ *for any $t > 0$,*

where $m(t)$ and $M(t)$ denote the mushy regions corresponding to the respective solutions (u, θ) and (U, Θ).

b) *If $D(\infty : u_0, \theta^0) > 0$ there exists a finite time t^* such that the mushy region disappears ($|m(t)| = 0$) for any $t \geq t^*$.*

c) *If $D(\infty : u_0, \theta^0) > \gamma$, with γ given by (13), the estimate $t^* \geq t_0^*$ holds, where t_0^* is defined by (14).* ∎

4. Remarks and sketch of the proofs.

A. We notice that in contrast with the radial case no assumption is made in Theorem 2 on the nature of $m(0)$ (the mushy region at $t=0$). So, for instance $m(0)$ may be a very irregular set with several connected components.

B. Assumptions (17) and (18) have a technical nature. One way to avoid them is to introduce $\theta_\#^0 \in W^{1,\infty}(0, \infty)$ be any nondecreasing function such that $0 \leq \theta_\#^0(t) \leq \theta^0(t)$ for any $t \geq 0$ and to define $\underline{u}_0(\cdot) = \min\{u_0(\cdot), \theta_\#^0(0)\}$. Then the conclusions of Theorem 2 remains true for general data u_0 and θ^0 by replacing the definition of (U, Θ) by the solution of the radial problem corresponding to the data $\underline{u}_0$ and $\theta_\#^0$.

C. Theorem 2 and the definition of D_∞ show that the behavior of the mushy region is influenced by an appropriate combination of the boundary temperature of the initial enthalpy. In particular the latter enters this combination as the integral of $\tilde{u}_0(\omega_N r^N) \varphi_N(r)$. We also remark that the weight function $\varphi_N(r)$ is peculiar to the boundary condition under consideration (so for the Neumann problem $\varphi_N(r)$, must be replaced by another suitable function $\psi_N(r)$, see [3]).

D. The main step of the proof of Theorem 1 is the equality

(21) $\displaystyle\int_0^{r_0} u(r, t) \varphi_N(r) dr = D(t : u_0, \theta^0)$

which is obtained by taking a suitable test function η in (8). Conclusion (i) follows easily from (21) arguing by contradiction. Conclusions (ii) and (iii) are shown by regularizing the problem and by using the structure of the mushy region $m_\varepsilon(t)=B(0,R_\varepsilon(t))$ of the regularized problem (here $R_\varepsilon(t)$ is known to be a strictly decreasing function of t).

E. The proof of Theorem 2 consists of several steps. The first and fundamental point is to prove the inequality

$$(22) \qquad \int_s^{|\Omega|} (\tilde{u}(\sigma,t))d\sigma \geq \int_s^{|\Omega|} (\tilde{U}(\sigma,t))d\sigma$$

for any $s\in[0,|\Omega|]$. That inequality is first obtained for a regularized problem (function ϕ is replaced by a Lipschitz sequence of functions ϕ_ε) and then by passing to the limit. The notion of relative rearrangement is used, as in [9], extending the approach of [6] to the case of time-depending boundary data. Finally conclusion (a) is shown using some arguments introduced in [4] (Theorem 1.28). Conclusions (b) and (c) comes easily from Theorem 1.

5. References.

[1] M. Bertsch, P. de Mottoni and L.A. Peletier: *"The Stefan problem with heating: appearance and disappearance of the mushy region"*. Trans. Amer. Math. Soc. 293 (1986), pp 677-691.

[2] J.I. Diaz, A. Fasano and A.M. Meirmanov: *"On the mushy region in multidimensional Stefan Problems with Dirichlet boundary conditions"*. To appear.

[3] J.I. Diaz, A. Fasano and M. Meirmanov: *"On the mushy region in multidimensional Stefan Problems with Neumann boundary conditions"*. To appear.

[4] J.I. Diaz: "Nonlinear partial differential equations and free boundaries. Vol I. Elliptic equations. Pitman Research Notes in Math. n^{0}106. Pitman, London, (1985).

[5] A. Fasano and M. Primicerio. *"General free-boundary problems for the heat equation III"*. J. Math. Anal. Appl. 59 (1977), pp. 1-14.

[6]   B. Gustafsson and J. Mossino: *"Isoperimetric inequalities for the Stefan problem"*. SIAM J. Math. Ana. 20 (1989) pp. 1095–1108.

[7]   A.M Meirmanov: "The Stefan problem". Novosibirsk, Nauka. (1986) (in russian). (An english translation will appear in Yolter de Gruenter ed).

[8]   A.M. Meirmanov and I.A. Kaliev. *"One-dimensional Stefan problem with an arbitrary initial enthalpy. Periodical solutions"*. In Free boundary problems: Applications and theory. Vol. III. A. Bossavit et al. eds. Pitman Research Notes in Math. n^{0}120 (1985), pp. 40–49.

[9]   J. Mossino and J.M. Rakotoson: *"Isoperimetric Inequalities in Parabolic Equations"*. Annali Scuola Normale Superiore di Pisa, 13 (1986), pp. 51–73.

[10] J.F. Rodrigues. *"The Stefan problem revisted"*. In Mathematical Models for Phase Change Problems. J.F. Rodrigues ed. Birkhäuser Verlag Basel (1989). pp. 129–190.

J.I. Diaz
Dpto Mat. Aplicada
Univ. Complutense Madrid
28040 Madrid SPAIN

A. Fasano
Dipto. Mat. "U.Dini"
Univ. Firenze
50134 Firenze, ITALY

A. Meirmanov
Lavrentyev Inst. Hydrodynamics
USSR Academy of Sciences,
630090 Novosibirsk, USSR

B GUSTAFSSON AND J MOSSINO
Isoperimetric inequalities for the Stefan problem

1. Introduction

We consider the Stefan problem in its simplest form and in an annular space geometry : find a pair (θ, h) of functions defined in $q = \omega \times (0, T)$ such that, in some weak sense,

(1.1)
$$\begin{cases} \dfrac{\partial h}{\partial t} - \Delta \theta = 0 \text{ in } q, \\ \theta = g \text{ on } \sigma = \partial \omega \times (0, T), \\ h_{|t=0} = h_0, \\ h \in a(\theta) \text{ a.e. in } q. \end{cases}$$

Here
· $\omega = \omega_0 \backslash \bar{\omega}_1$; ω_0, ω_1 bounded regular domains in $\mathbb{R}^N$ ($N \geq 2$); $\bar{\omega}_1 \subset \omega_0$;
· g constant on each of $\sigma_j = \gamma_j \times (0, T) = \partial \omega_j \times (0, T)$ $(j = 0, 1)$, let us say

$$g = \begin{cases} 0 \text{ on } \sigma_0 \\ 1 \text{ on } \sigma_1; \end{cases}$$

· a is a strictly monotone graph in $\mathbb{R}^2$:

(1.2)
$$a(\theta) = \begin{cases} \alpha_0(\theta - \lambda) - \alpha \text{ for } \theta < \lambda, \\ [-\alpha, 0] \qquad \text{for } \theta = \lambda, \\ \alpha_1(\theta - \lambda) \qquad \text{for } \theta > \lambda, \end{cases}$$

$\alpha, \alpha_0, \alpha_1$ positive constants, $\lambda \in [0, 1]$;
· $h_0 \in L^\infty(\omega)$ satisfies an extra condition (see (1.6), (1.7) below), which essentially means that $\theta_0 = b(h_0) = a^{-1}(h_0)$ belongs to $H^1(\omega)$ and $0 \leq \theta_0 \leq 1$.

Our boundary and initial data, g and h_0, are such that $0 \leq \theta \leq 1$ in all q, by the maximum principle. If $\lambda = 0$ in (1.2), the temperature θ in the solid phase (the latter generally defined as the region where $h \leq -\alpha$ ($h = -\alpha$ if $\lambda = 0$)) therefore must be constantly equal to zero. Similarly, if $\lambda = 1$, the temperature in the liquid phase $\{h \geq 0\}$ ($h = 0$ here) is constantly equal to 1. Thus, for these extreme cases, in practice we have a one-phase Stefan problem, while for $0 < \lambda < 1$, the problem really is a two-phase problem.

One standard way of making (1.1) precise is to say that (θ, h) is a weak solution of (1.1) if

(1.3)
$$
\begin{cases}
\theta \in L^\infty(q), \ h \in L^\infty(q), \ h \in a(\theta) \ \text{a.e. in } q, \\
\displaystyle\int\int_q \left(h \frac{\partial \varphi}{\partial t} + \theta \Delta \varphi \right) dx dt = \\
\displaystyle= \int\int_\sigma g \frac{\partial \varphi}{\partial \nu} d\gamma dt - \int_\omega h_0(x) \varphi(x, 0) dx
\end{cases}
$$

for every "test function" $\varphi \in C^1(\bar{q})$ satisfying $(\partial^2 \varphi / \partial x_i \partial x_j) \in C(\bar{q})$ and $\varphi = 0$ on $\sigma \cup (\omega \times \{T\})$ (see, e.g. [1] or [2]).

One can obtain the weak solution as a limit as $\epsilon \to 0$ ($\epsilon > 0$) of the classical solutions $(\theta_\epsilon, h_\epsilon)$ of some regularized problems $(1.1)_\epsilon$ where
· a is replaced by a_ϵ, single-valued smooth function with

(1.4)
$$a'_\epsilon \geq \delta > 0 \ (\delta \ \text{independent of } \epsilon)$$

(1.5)
$$b_\epsilon = a_\epsilon^{-1} \ \text{converges uniformly to } b = a^{-1};$$

· h_0 is replaced by smooth functions h_{ϵ_0}

(1.6)
$$h_{\epsilon_0} \to h_0 \ \text{in } L^1(\omega),$$

and $\theta_{\epsilon_0} = b_\epsilon(h_{\epsilon_0})$ satisfies

(1.7)
$$
\begin{cases}
0 \leq \theta_{\epsilon_0} \leq 1 \ \text{in } \omega, \ \theta_{\epsilon_0|\partial\omega} = g, \\
\displaystyle\int_\omega |\nabla \theta_{\epsilon_0}|^2 \ dx \ \text{is bounded independently of } \epsilon, \text{as } \epsilon \to 0.
\end{cases}
$$

(see [7], [1], [2]).

We give below isoperimetric inequalities for the weak solution of (1.1). They are not standard, since the problem is multivalued, and the domain is doubly connected.

2. Statements of results.

Our isomerimetric inequalities are inequalities between certain quantities for the problem (1.1) (or $(1.1)_\epsilon$) and the corresponding quantities for certain "symmetrized problems" ($(\widetilde{1.1})$ and $(\widetilde{1.1})_\epsilon$ below). Before describing these symmetrized problems, we must introduce some general definitions and notations (see e.g. [5]).

· $| E |$ denotes the volume (Lebesgue measure) of the measurable set E in $\mathbb{R}^N$;
· Ω_j $(j = 0, 1)$ denotes the open balls in $\mathbb{R}^N$ centered at the origin, such that $| \Omega_j | = | \omega_j |$; $\Omega = \Omega_0 \backslash \bar{\Omega}_1$; $Q = \Omega \times (0, T)$, $\Gamma_j = \partial \Omega_j$;
· f_* denotes the decreasing rearrangement of a measurable function defined in ω :

$$\forall s \in [0, | \omega |], \ f_*(s) = \text{Inf} \ \{t \in \mathbb{R}, \ | \{f > t\} | \leq s\},$$

while $\underset{\sim}{f}$ denotes the rearrangement of f, defined in Ω, that decreases along radii ; if f is defined in q, we consider its rearrangements with respect to the space variable.

Now, the symmetrized problem corresponding to (1.1) is

$$(\widetilde{1.1}) \qquad \begin{cases} \dfrac{\partial H}{\partial t} - \Delta \Theta = 0 \text{ in } Q, \\[2mm] \Theta = G \text{ on } \Sigma = \partial \Omega \times (0, T), \\[2mm] H_{|t=0} = \underset{\sim}{h_0} \\[2mm] H \in a(\Theta) \text{ a.e. in } Q, \end{cases}$$

where $G = 0$ on $\Sigma_0 = \Gamma_0 \times (0, T)$, $G = 1$ on $\Sigma_1 = \Gamma_1 \times (0, T)$. Similarly the symmetrized problem corresponding to $(1.1)_\epsilon$ is $(\widetilde{1.1})_\epsilon$.

Our main technical tool is the following result

THEOREM 1. *For weak solutions of* (1.1) *and* $(\widetilde{1.1})$ (*see* (1.3))

$$(2.1) \qquad \begin{aligned} \int_0^s h_*(\sigma, t)d\sigma - \int \int_{\sigma_1} \frac{\partial \theta}{\partial \nu} d\gamma d\tau &\leq \int_0^s H_*(\sigma, t)d\sigma \\ &- \int \int_{\Sigma_1} \frac{\partial \Theta}{\partial \nu} d\gamma d\tau \end{aligned}$$

for every (s,t) in $[0, | \Omega |] \times [0, T]$, where ν is the outward normal.

Inequality (2.1) follows for the corresponding $(2.1)_\epsilon$ for classical solutions of $(1.1)_\epsilon$ and $(\widetilde{1.1})_\epsilon$, and $(2.1)_\epsilon$ relies on the techniques developed in [6] for linear parabolic problems.

We now give some other isoperimetric inequalities that have more physical significance and therefore can be viewed as the main results. They are all simple consequences of Theorem 1.

First, in the particular cases $s = 0$ and $s = | \omega |$, (2.1) reduces to the comparison of the heat flows through the boundaries

$$(2.2) \qquad \int \int_{\sigma_1} \frac{\partial \theta}{\partial \nu} d\gamma d\tau \geq \int \int_{\Sigma_1} \frac{\partial \Theta}{\partial \nu} d\gamma d\tau \ (\geq 0),$$

$$(2.3) \qquad -\int \int_{\sigma_0} \frac{\partial \theta}{\partial \nu} d\gamma d\tau \geq -\int \int_{\Sigma_0} \frac{\partial \Theta}{\partial \nu} d\gamma d\tau \ (\geq 0).$$

Moreover, for the "regular" solutions

$$(2.4) \qquad -\int_0^t \int_{\gamma(s,\tau)} \frac{\partial \theta}{\partial \nu} d\gamma d\tau \geq -\int_0^t \int_{\Gamma(s,\tau)} \frac{\partial \Theta}{\partial \nu} d\gamma d\tau \ (\geq 0)$$

with $\gamma(s,t) = \{x \in \omega, h(x,t) = h_*(s,t)\}$, Γ defined similarly. The members of (2.4) have the physical interpretation of being the total heat flows during the time interval $(0, t)$ from the warmest parts of volume s of ω and Ω, respectively into the complementary colder parts.

In the case of a one-phase Stefan problem, let us say with $\lambda = 0$, the quantity

$$(2.5) \qquad t' = \sup \ \{t \in [0, T], \int_0^\tau \int_{\gamma_0} \frac{\partial \theta}{\partial \nu} d\gamma d\tau' = 0, \ a.e. \tau \in (0, t)\}$$

can be interpreted as the first instant at which the liquid phase reaches γ_0 (when $\lambda > 0, t' = 0$). With the similar definition for $(\widetilde{1.1})$, T' also appears to be the time at which all the solid has melted.

THEOREM 2. *With the above definitions*

$$(2.5) \qquad \qquad t' \leq T',$$

$(2.6) \ \int_s^{|\omega|} h_*(\sigma, t) d\sigma \geq \int_s^{|\omega|} H_*(\sigma, t) d\sigma, a.e. t \in (0, t'), \forall s \in [0, | \omega |]$

$(2.7) \qquad | \{x \in \omega, h(x,t) = -\alpha\} | \leq | \{x \in \Omega, H(x,t) = -\alpha\} |, a.e. t \in (0, t').$

133

The latter inequality expresses that the volume of the solid remains greatest in spherical geometry (up to time t').

Some more isoperimetric inequalities, and the detailed proofs are given in [4]. The first results were announced in [3].

References

[1] A. Friedman, The Stefan problem in several space variables, *Trans. Amer. Math. Soc. 139* (1968), 51-87.

[2] A. Friedmand, *Variational Principles and Free Boundary Problems*, John Wiley, New York, 1982.

[3] B. Gustafsson and J. Mossino, Quelques inégalités isopérimétriques pour le problème de Stefan, *C. R. Acad. Sci. Paris, Série I*, (1987), 669-672.

[4] B. Gustafsson and J. Mossino, Isoperimetric inequalities for the Stefan problem, *SIAM J. Math. Anal. 20*, n° 5, (1989), 1095-1108.

[5] J. Mossino, *Inégalités Isopérimétriques et Applications en Physique*, Hermann, Paris, 1984.

[6] J. Mossino and J.M. Rakotoson, Isoperimetric inequalities in parabolic equations, *Ann. Scuola Norm. Sup. Pisa, Cl. Sci. 4, 13* (1986), 51-73.

[7] O.A. Oleinik, On the equations of the type of nonstationary filtration, *Dokl. Akad. Nauk. SSSR 113* (1957), 1210-1213.

B. Gustafsson

Department of Mathematics, Royal Institute of Technology, S-10044 Stockholm, Sweden.

J. Mossino

Département de Mathématiques, Bâtiment 425, CNRS et Université Paris-Sud, 91405 Orsay, France.

E RADKEVICH

The Gibbs–Thomson effect and existence conditions for classical solutions for the modified Stefan problem

The aim of this paper is to obtain, on a small time interval, the existence conditions for a classical solution, i.e. a solution to a sharp interface modified Stefan problem, which allows for surface tension and kinetic undercooling in the domain $\Omega \subset \mathbb{R}^n$, $n \geq 2$. The problem is to find a function $u_j(x,t)$, $j=1,2$ (i.e. the temperature) and a surface $\Gamma(t) \subset \Omega, \forall t \subset [0,T]$ (i.e. the free boundary) such that

$$L_j u_j \equiv \left(\frac{\partial}{\partial t} - \nabla_x (A_j(x,t) \nabla_x) \right) u_j = f_j \text{ in } \Omega_j(t), \forall t \ \varepsilon \ [0,T], \tag{1}$$

where the domain $\Omega_1(t) \subset \Omega$ has the boundary $\partial\Omega_1(t) = \Gamma(t)$, and $\Omega_2(t) = \Omega \backslash \overline{\Omega}_1(t)$, $\forall t \subseteq [0,T]$;

$$\ell \ V_n = [A \nabla_x u \cdot n_t]_-^+ \text{ on } \Gamma(t); \tag{2}$$

$$u_j = -\sigma(x,t)k(x,t) - \alpha(x,t)\sigma(x,t)V_n \text{ on } \Gamma(t), j=1,2 \tag{3}$$

$\forall t \subseteq [0,T]$, and

$$u_j|_{t=0} = j_{0j} \text{ in } \Omega_j(0); \tag{4}$$

$$u_2 = g_D \text{ or } \frac{\partial}{\partial v_2} u_2 = g_N \text{ on } (\partial\Omega)_T , \tag{5}$$

where, for any set $Q \subset \mathbb{R}^k$, $k \geq 0$, $Q_T = Q \times [0,T]$, $\frac{\partial}{\partial v_2}$ denotes the derivative with respect to the operator L_2 conormal to $(\partial\Omega)_T$. Here ℓ is the latent heat, V_n is the

135

(normal) velocity of the interface $\Gamma(t)$ exterior to $\Omega_1(t)$; the symbol $[\cdot]\pm$ on the right hand side of (2) designates the values difference between $A_2\nabla_x u_2 n_t$ and $A_1\nabla_x u_1 n_t$ on the free boundary in the direction from solid to liquid (i.e. the conormal derivative jump on the free boundary in the direction from solid to liquid), where n_t is the unit normal to $\Gamma(t)$ exterior to Ω_1; $\sigma(x,t)>0$ is the surface tension, $\alpha(x,t) > 0$ is the relaxation parameter; $k(x,t)$ is the local sum of the principal interface curvatures at the point $(x,t)\epsilon\Gamma(t)$ (see [1]).

Now for convenience we make the following assumptions: the coefficients $A_{jki}(x,t)$ of the matrix $A_j(x,t)$ and also $\sigma(x,t)$ and $\alpha(x,t)$ belong to $C^{i_0,i_0/2}(\overline{\Omega}_{T_0})$ $T_0 > 0$, also we assume

$$c_0 \leq \sum_{k,i} A_{jki}(x,t)\xi_k\xi_i \leq c_0^{-1}, \ \forall \ (x,t) \ \epsilon \ \Omega_{T_0}, \ \xi \ \epsilon \ \mathbf{R}^n, \ |\xi| = 1.$$

and that the boundaries $\partial\Omega$ and $\Gamma(0)$ belong to C^{i_0}, where i_0 is sufficiently large, and $\partial\Omega\cap\Gamma(0) = \varnothing$.

We will assume the reducibility of matrices A_j, $j=i, 2$, i.e. the existence at any point $\wp\subset\Gamma(0)$ of a ball $U(\wp)\subset\Omega$ with a radius $r(\wp)\geq r_0>0$ and a differmorphism $T_\wp: U(\wp)\rightarrow V$, where V is a neighbourhood of the origin in $\mathbf{R}^n_z$ with unit Jacobi matrix such that, in new coordinates $z=T_\wp(x)$, the operator (1) has the following representation:

$$L_j u_j \equiv (\frac{\partial}{\partial t} - \nabla_z (A_j'(z,t)\nabla_z) - a_j'(z,t)\nabla_z)u_j,$$

where the coefficients A'_{jki} of matrix A'_j are represented in the form:

$$A_{jki}'(z,t) = a_j(\wp)\delta_{ki} + b_{jki}(z,t), b_{jki}(\wp,0) = 0,$$

and $\delta_{ki} = 0$ if $k \neq i$ and $\delta_{kk} = 1$.

The attention of many physicists and mathematicians to this problem has recently been drawn by the following observation: in the first place, this model describes the needle crystal growth (see [2]); second, this model is obtained for the asymptotic behaviour of the phase field method

$$u_t + \frac{\ell}{2}\phi_t = K\nabla_x^2 u, \tag{6}$$

$$\alpha\xi^2\phi_t = \xi^2\nabla_x^2\phi + \frac{1}{2a}(\phi - \phi^3) + 2u. \tag{7}$$

As is shown in [3] for $\xi a^{-1/2} = \frac{3}{2}\sigma = O(1)$, $\varepsilon = \xi^{-1}a^{1/2} \to 0$, the problem (6), (7) is an approximation of the modified Stefan problem (1) - (5). When $\xi a^{-1/2} \to 0$, $\xi^{-1}a^{1/2}\alpha \to 0$ it is an approximation of the classical Stefan problem.

Let us formulate the main result of this paper.

Theorem. Let the manifolds $\partial\Omega$, $\Gamma(0)$ and the coefficients of matrix $A_j(x,t)$ satisfy the above conditions. Let $k_{0j}(\wp)$, $j=1,....,n-1$, be the principle curvature at any point $\wp \subset \Gamma(0)$ and

$$\frac{\partial}{\partial n_0}\phi_{0j}(\wp) > \sigma \sum_{s=1}^{n-1} k_{0s}^2(\wp), \forall \wp \varepsilon \Gamma(0), j=1,2, \tag{8}$$

where n_0 is unit normal to $\Gamma(0)$ to be exterior to $\Omega_1(0)$.

Then, for any initial-boundary values such that:

$$f_j \varepsilon C^{i_0-2, i_0-2/2}(\Omega_{T_0}), g\varepsilon C^{i_0-\gamma, i_0-\gamma/2}((\partial\Omega)_{T_0}),$$

where $\gamma = 0$ for the D problem and $\gamma = 1$ for N problem (depending on the condition (3)), $\phi \subset C^{i_0}(\Omega_0)$, and for which the compatibility conditions are correct to the order $[i_0/2]$ for $t = 0$, the classical solution of the problem (1) - (5) exists on a sufficiently small time interval $[0,T]$, $T < T_0$. Moreover, the surface $\Gamma = \bigcup_{t=0}^{T} \Gamma(t)$ belongs to

$$C^{i,i/2}, u_j \varepsilon C^{i,i/2} \left(\bigcup_{t=0}^{T} \Omega_j(t) \right), \ j=1,2, \text{ and } i_0 - 3 > i > 4, \ [i] \neq 1$$

In order to prove this theorem, the methods used by the author for the existence proof of the classical solution of the Stefan, Verigin - Muscat problems and the system of binary alloy solidification (see [4-5]) can be applied. The methods used in these papers are as follows.

First of all, one transfers from the problem (1) - (5) in the time domains Ω_j (t) to the problem in the domains Ω_j (0), by applying the Hanzawa transformation $g_{\rho,T}$ [6]. To do this we define in a neighbourhood $N_0 \subset \Omega$ of $\Gamma(0)$ the local coordinate system $x = x(\lambda_n, \omega)$, where $\lambda_n(x)$ is the distance from the point x to $\Gamma(0)$ and $\omega(x)$ is the local coordinate system on $\Gamma(0)$ corresponding to the projection $P(x)$ of the point x on $\Gamma(0)$. For any function:

$$\rho \varepsilon B_i(\Gamma(0)_T) = \{\rho \varepsilon C^{i,i/2}(\Gamma(0)_T), \|\rho; C^{2,1}(\Gamma(0)_T)\| \leq B; \rho|_{t=0} = 0\}$$

we define
$$g_{\rho,T} : (N_0)_T \rightarrow \Omega_T$$

such that
$$g_{\rho,T}(t)(x,t) = (x(\lambda_n + \chi(\lambda_n)\rho(\overline{\omega},t); \omega),t),$$

where $\chi \varepsilon C_0^{00}(\Omega), 0 \leq \chi \leq 1, \chi = 0$ outside N_0 and $\chi = 1$ in a neighbourhood $N_0' \subset N_0$ for $\Gamma(0)$. Outside $(N_0)_T$, the transformation $g_{\rho,T}$ coincides with the identity transformation. It is easy to check that $g_{\rho,T}$ is the differmorphism if:

$$|1 + \chi'(\lambda_n)\rho(\omega,t)| \geq 1 - B|\chi'| \geq 1/2 \text{ in } (N_0)_T.$$

We define the nonlinear operator on the functions $\rho \subset B_1(\Gamma(0)_T)$ such that

$$\beta(\rho) = \frac{\partial}{\partial t}\rho + \sum_{j=1}^{2} (-1)^j A_j \nabla_x u_{j\rho} \nabla_x h_\rho, \ h_\rho = \lambda_n - \rho, \tag{9}$$

where $u_{j\rho}$, $j=1,2$ is the solution of the problem (1), (3) - (5) in the domains $\Omega_{\rho,T}(t) = g_{\rho,T}(t)(\Omega_j(0))$ with interface $\Gamma_\rho(t) = g_{\rho,T}(t)(\Gamma(0))$. Taking the compatibility conditions of (1) - (5) for t=0 into account we can define uniquely the following derivatives:

$$R_k = D_t^k \rho|_{t=0}, \ U_{jk} = D_t^k u_j|_{t=0}, \ k=0,....,[i_0/2].$$

This permits us to use the trace theorem to construct the function

$\rho_0 \subset C^{i_0, i_0/2}(\Gamma(0)_{T_0})$ such that $D_t^k \rho_0|_{t=0} = R_k, k=0....,[i_0/2]$, and then to solve the problem (1), (3) - (5) in the domains $\Omega_{j\rho}(t)$. By virtue of the compatibility conditions these solutions

$$u_{j\rho_0} \ \varepsilon \ C^{i_0-2, i_0-2/2}\left(\overset{T}{\underset{t=0}{U}} \ \Omega_{j\rho_0}(t) \right)$$

and the function ρ_0 determine the initial approximation of the following problem:

$$\beta(\rho) = 0, \ \rho \varepsilon B_1(\Gamma(0)_T) \tag{10}$$

Solving (10) we can obtain the classical solution $\{u_{1\rho}, \ u_{2\rho}, \ \Gamma_\rho(t)\}$ of the problem (1) - (5). Really, in view of the compatibility conditions and the structure, we have:

$$D_t^k \ \beta \ (\rho_0)|_{t=0} \ = \ 0, \ j=0,.....[i_0 -3/2].$$

Further, taking into account $v_{j\rho} = u_{j\rho} * g_{\rho,T}$ and rewriting the equations (1) - (5) as the equations for the function $v_{j\rho}$ in the domains $(\Omega_j(0))_T$ with the boundary equations (2), (3) on the surface $\Gamma(0)_T$, $T < T_0$ (see [4-8]), it is easy to write the Frechet derivative for the operator $\beta(\rho)$.

To use the method of successive Newton approximations, it is necessary to obtain the invertibility conditions for the Frechet derivative for the initial approximation ρ_0. Here it is necessary to use the classical method of freezing the coefficients which permits us to reduce the problem to the investigation for any point $\wp \subset \Gamma(0)$, the single solvability conditions for the following model problems (see [4-7]):

$$(\frac{\partial}{\partial t} - a_j(\wp)\nabla_z^2) \, W_j = 0 \ \text{in} \ (R_j^n)_T, \, j=1,2, \tag{11}$$

where

$$R_j^{\ n} = \{z \varepsilon R^{\ n}, (-1)^j z_n > 0\}, \quad W_j|_{t=0} = 0; \tag{12}$$

$$\frac{\partial}{\partial t} \delta\rho = -\sum_{j=1}^{2} (-1)^j a_j(\wp) D_{z_n} W_j + G_1 \quad \text{on} \quad (\Gamma_0)_T, \tag{13}$$

where $\Gamma_0 = \{z \varepsilon R^{\ n}, z_n = 0\}$,

$$W_j = \sigma(\wp) \nabla_{z'}^2 \delta\rho + \left(\sum_{k=1}^{n-1} \sigma(\wp) \kappa_{0k}^2(\wp) - \frac{\partial}{\partial n_0} \phi_{0j}(\wp) \right) \delta\rho \tag{14}$$

$$- \alpha(\wp)\sigma(\wp) \frac{\partial}{\partial t} \delta\rho + G_2 \quad \text{on} \quad (\Gamma_0)_T,$$

where the functions G_j have the compact supports such that

$$G_1 \varepsilon C^{0i,i/2}((\Gamma_0)_T), \quad G_2 \varepsilon C^{0i-1,(i-1)/2}((\Gamma_0)_T)$$

and the integer $i > 2$. Here:

$$C^{0_{s,s/2}} = \{g \varepsilon C^{s,s/2}, D_t^k g|_{t=0}, k = 0, \dots, [s/2]\}.$$

By expressing $\delta\rho$ from the equations (11)-(14), i.e. reducing the problem on the boundary, we obtain the Cauchy problem which is the analog of the Schapiro-Lopatinski conditions (see [4,6,7]):

$$\mathfrak{C}(D_t, D_{z'}) \delta\rho = F \quad \text{on} \quad (\Gamma_0)_T, \ \delta\rho|_{t=0} = 0. \tag{15}$$

where the symbol $\widetilde{\mathfrak{C}}$ of the operator $\mathfrak{C}$ has the form:

140

$$\mathfrak{C}(ip, i\zeta') = ip + \sum_{j=1}^{2} (a_j(\mathbf{\Phi}))^{1/2} (ip + a_j(\mathbf{\Phi})\zeta'^2)^{1/2}$$

(16)

$$[\sigma(\mathbf{\Phi})\zeta'^2 + (D_{\eta_n}\phi_{0j}(\mathbf{\Phi}) - \sigma(\mathbf{\Phi})\sum_{k=1}^{n-1}\kappa_{ok}^2(\mathbf{\Phi}))$$

$$+ \alpha\sigma(\mathbf{\Phi})ip], \quad \zeta' = (\zeta_1,....,\zeta_{n-1})$$

If the condition (8) is correct and for any $\sigma \geq 0$, $\alpha \geq 0$, the symbol $\mathfrak{C}$ satisfies the conditions of Lemma 1 [4] (see the chapter 4 [4], i.e. it is parabolic, and hence the fundamental solution $G(z',t)$ of the Cauchy problem exists, so in the case when $\sigma \geq 0$, $\alpha \geq 0$, we have

$$D_t^\alpha D_{z'}^\beta G(z',t)| \leq \text{Const } t^{-(n-2+2d+|\beta|)/2} \exp[-\omega_0|z'|^2/t], \omega_0 = \text{Const} > 0.$$

From here it is easy to prove for any function G_j , with compact supports, the existence of a unique solution of the problem (11)-(14), which in the case $\sigma > 0$, $\alpha > 0$ satisfies the following inequality:

$$\sum_{j=1}^{2} \|W_j; C^{0i,i/2}\left((R_j^n)_T\right)\| + \|\delta\rho; C^0C^{i+2,i+2/2}((\Gamma_0)_T)\|$$

$$\leq \text{Const}[\|G_1; C^{0i,i/2}((\Gamma_0)_T)\| + \|G_2; C^0C^{i-1,(i-1)/2}((\Gamma_0)_T)\|]$$

From these estimates we obtain the structure needed to implement the Frechet derivative regularization, i.e. the proof of its reversibility. Further, using the method of Newton's successive approximations, one can prove the existence of a classical solution of the problem (1) - (5) on the small time interval [0,T], $T < T_0$.

REFERENCES

1.   G. Caginalp, "Mathematical models of phase boundaries" Symposium on Material Instabilities in Continuum Mechanics, Heriot-Watt Univ. (1985-86) 35-50. Ed. J. Ball Oxford Science Publ. (1988)

2.   B. Caroli, C. Caroli, C. Misbah, B. Roulet - J. Physique. (1987) v. 48, p. 547.

3.   G. Caginelp. "The role of microscopic anisotropy in the macroscopic behaviour of a phase boundary" Annals. of Physics (1986) v. 172, p. 136-155.

4.   E. Radkevich, A. Melikulov, "Boundary problems with free boundaries", FAN, Tashkent (1988) (in Russian).

5.   E. Radkevich, "On the operator clusters contact Stefan problem", Matematicheskie zametki (1990) v. 47, N2, p. 89-100 (in Russian).

6.   E. Radkevich, "On solvability general unstationary problems with free boundary", Doklady AN SSSR (1986) v. 281, N5, p. 1091-1099 (in Russian).

7.   E. Radkevich, A. Melikulov, "Problems with free boundary", FAN. Tashkent (1991) (to appear) (in Russian).

8.   Ei-Ichi Hanzawa, "Classical solutions of the Stefan problem", Tohoku Math. J. (1981) v. 33, N3, p. 297-335.

E. Radkevich
Department of Mathematics
Moscow Institute of Radio Engineering Electronics
and Automatic (MIREA)

J F RODRIGUES
Stability of the free boundary in a two-phase continuous casting Stefan problem

In this note we study the regularity and the stability of the free boundary of a two-phase Stefan problem with convection arising in a simple mathematical model for the continuous casting solidification. For a two-dimensional evolutive model, under certain monotonicity conditions, we show that the free boundary is given, for each instant, by a Lipschitz continuous curve, which varies continuously with time. We establish its continuous dependence with respect to the data and we show the asymptotic convergence towards the steady-state solution as $t \to \infty$.

We represent the half longitudinal section, in a fixed frame, by the rectangle $\Omega =]0, \ell [\times]0, L[\subset \mathbb{R}^2$, $(\ell, L>0)$ and the constant extraction velocity by $V = \nu \, Z \, (\nu >0)$, so that the temperature $\vartheta = \vartheta(x,z,t)$, in $Q_T = \Omega \times]0,T[$, and the free boundary $\Phi = \{(x,z,t) \in Q_T : \psi(x,z,t) = 0\}$ satisfy

$$b(\vartheta)(\partial_t \vartheta + \nu \, \partial_z \vartheta) = \Delta \vartheta \quad \text{in } Q_T \backslash \Phi = \{\vartheta<0\} \cup \{\vartheta>0\} \tag{1}$$

$$\vartheta = 0 \quad \text{and} \quad \lambda(\partial_t \psi + \nu \, \partial_z \psi) = [[\nabla \vartheta . \nabla \psi]] \quad \text{on } \Phi \tag{2}$$

for a given $\lambda >0$ (the latent heat) and a real function b, which we assume strictly positive and locally bounded (here $[[.]]$ denotes the jump across the free boundary). The associated boundary and initial conditions are for $t > 0$ and $t=0$:

$$\vartheta(x, 0, t) = M > 0 \quad \text{and} \quad \vartheta(x, L, t) = \mu < 0, \quad \text{for } 0 < x < \ell, \tag{3}$$

$$\partial_x \vartheta(0, z, t) = 0 \quad \text{and} \quad \partial_x \vartheta(\ell, z, t) = g(z, t, \vartheta(\ell, z, t)) \quad \text{for } 0 < z < L, \tag{4}$$

$$\vartheta(x,z,0) = \vartheta_0(x, z), \quad (x,z) \in \Omega. \tag{5}$$

The Dirichlet conditions in (3) imply that the left boundary is completely at liquid phase, while the right side is totally solidified. The first condition in (4) is just a symmetry condition, while the second one traduces a nonlinear flux representing the lateral cooling.

This problem with $\nu =0$ (i.e., without convection) is largely studied (see, for instance, [D],[M] or [R1] and their references). The case $\nu >0$ has been recently studied in [RY], where it was proved that there exists a unique weak solution $(\vartheta,\eta) \in L^2(0,T;H^1(\Omega)) \times L^2(Q_T)$ to the problem (1-5),

$$\eta \in \beta(\vartheta) \quad \text{a.e. in } Q_T, \quad \vartheta = M \text{ on } \Sigma_T^0 \text{ and } \vartheta = \mu \text{ on } \Sigma_T^L, \quad \text{and} \tag{6}$$

143

$$- \int_{Q_T} \eta (\partial_t w + \nu \, \partial_z w) + \int_{Q_T} \nabla \vartheta . \nabla w = \int_{\Sigma^N_T} g(\vartheta) w + \int_{\Omega} \eta_0 w(0), \qquad \forall w \in W_0 \qquad (7)$$

which, in addition, satisfies $\vartheta \in C^0(\overline{Q}_T)$ and $\mu \leqslant \vartheta(x,z,t) \leqslant M$ for $(x,z,t) \in Q_T$.

Here we set $\Sigma^j_T = \Gamma_j x]0,T[$, $j=0$, L, N, $\Gamma_j =]0,\ell [x\{j\}$, $j=0,L$, $\Gamma_N = \{\ell \}x]0,L[$ and the space of test functions is given by $W_0 = \{ w \in H^1(Q_T): w(T)=0$ in Ω , $w=0$ on $\Sigma^0_T \cup \Sigma^L_T\}$, β denotes the maximal monotone graph $\beta(\Theta)=\int_0^\Theta b(\sigma)\, d\sigma$ if $\Theta <$ 0, $\beta(\Theta)=[0,\lambda]$ if $\Theta = 0$, and $\beta(\Theta)=\lambda + \int_0^\Theta b(\sigma)\, d\sigma$, if $\Theta > 0$; μ, M and G are given constants and g: $\Sigma^N_T \times \mathbb{R} \to \mathbb{R}$ is a measurable function, such that, $g(\vartheta)=g(z,t,\vartheta)$ is Lipschitz continuous and monotone non-increasing in ϑ, for a.e. $(z,t) \in \Sigma^N_T$, satisfying

$G \leqslant g(z,t,\vartheta) \leqslant 0$ if $\vartheta > M$ and $g(z,t,\vartheta)=0$ if $\vartheta \leqslant \mu$, a.e. $(z,t) \in \Sigma^N_T$

$\eta_0 \in \beta(\vartheta_0)$ a.e. in Ω, for some $\vartheta_0 \in C^0(\overline{\Omega})$,

$\mu < 0 < M$ and $\mu = \vartheta_0(x, L) \leqslant \vartheta_0(x, z) \leqslant \vartheta_0(x, 0)=M$, $(x,z) \in \overline{\Omega}$.

The sign restriction $g \leqslant 0$ corresponds to a cooling process. It may replaced, in the linear case (with $0 \leqslant \gamma(z,t) \leqslant \gamma^*$ and $\mu \leqslant g_0(z,t) \leqslant M$) by

$$g(z,t,\vartheta) = \gamma(z,t)[g_0(z,t) - \vartheta], \quad (z,t) \in \Sigma^N_T, \vartheta \in \mathbb{R}. \qquad (8)$$

Using a suitable application of the maximum principle we have:

Proposition 1[R3]: In addition, to the conditions above, assume

$$\partial_z g \leqslant 0 \quad \text{and} \quad \partial_t g + \nu \partial_z g \leqslant 0 \quad \text{in } \mathcal{D}'(\Sigma^N_T \times \mathbb{R}) \qquad (9)$$

$$\partial_x \vartheta_0 \leqslant 0, \quad \partial_z \vartheta_0 \leqslant 0 \quad \text{and} \quad \Delta \vartheta_0 \leqslant 0 \quad \text{in } \mathcal{D}'(\Omega). \qquad (10)$$

Then we have

$$\partial_x \vartheta \leqslant 0, \quad \partial_z \vartheta \leqslant 0 \quad \text{and} \quad \partial_t \vartheta + \nu \partial_z \vartheta \leqslant 0 \quad \text{in } Q_T. \qquad (11)$$

Remark 1: In the case of a linear cooling given by (8), the sign restriction $g(\vartheta) \leqslant 0$ requires $g_0 \leqslant \vartheta$ on Σ^N_T, which can be guaranteed, for instance, if we assume the following set of restrictions: $g_0 \leqslant \vartheta_0$, $g_0(0+,t) \leqslant M$, $g_0(L-,t)=\mu$ and $\exists h=h(z,t) \in \beta(g_0(z,t))$ such that $\partial_t h + \nu \partial_t h \leqslant \partial^2_z g_0$ in $\mathcal{D}'(]0,L[\times]0,T[)$. These conditions imply that g_0 is a subsolution to ϑ (see [RY], and [R2] for the stationary case). On the other hand, the restriction (9) corresponds now to impose, respectively, $\partial_z \gamma \gtrless 0$, $\partial_z g_0 \lessgtr 0$ and $\partial_t \gamma + \nu \partial_z \gamma \gtrless 0$, $\partial_t g_{00} + \nu \partial_z g_0 \lessgtr 0$.

These monotonicity properties of the temperature can be used in order to show that the free boundary $\Phi = \{(x,z,t) \in \overline{Q_T} : \vartheta(x,z,t) = 0\}$, corresponding to (6)-(7) and separating the liquid region $\{\vartheta > 0\}$ from the solid one $\{\vartheta < 0\}$, may be given, at each time $t > 0$, by a graph

$$\Phi(t) = \{(\xi,\zeta) \in \Omega : \zeta = \phi(\xi,t)\} \tag{12}$$

in the rotated coordinate system $\xi = (z - x)/\sqrt{2}$ and $\zeta = (x + z)/\sqrt{2}$, where the function ϕ is Lipschitz continuous in ξ and continuous in t. In addition, if for each $t \in [0,T]$ we extend $\phi(\bullet,t)$ by continuity to $I = [-\ell/\sqrt{2}, L/\sqrt{2}]$, through the lines $\zeta = \xi$ and $\zeta = \xi + \sqrt{2}\ell$, we have the following additonal regularity for ϕ:

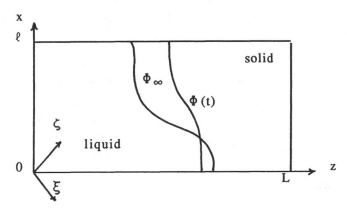

Theorem 1[R3]: The free boundary is given by the continuous graph (12) in $(\xi,\zeta,t) \in \overline{Q_T}$. Moreover $\phi \in C^0([0,T];C^{0,\alpha}(I)) \cap L^\infty(0,T;C^{0,1}(I))$, $0 \leqslant \alpha < 1$.

The non-existence of a mushy region, i.e. the fact that $meas\{\vartheta = 0\} = 0$ implies the constitutive relation

$$\beta(\vartheta) = \int_0^\vartheta b(\sigma)d\sigma + \lambda \, \chi_{\{\vartheta > 0\}},$$

where $\chi_{\{\vartheta > 0\}}$ denotes the characteristic function of the set $\{\vartheta > 0\}$. This relation is satisfied almost everywhere and, combined with a energy stability estimate and the regularity of the free boundary, allow us to deduce quantitative properties of continuous dependence of the interface with respect to the data (see also [R1]).

Remark 2: We observe that this non-existence of "mushy region" is due to (11) and it corresponds to a monotone solidification process with respect to the moving frame at constant velocity $v\,Z$. In the moving frame this is similar to the monotonicity restriction of [N]. However, we do not need here the uniform Lipschitz property for the temperature as in [N].

Let ϑ, ϕ and $\hat{\vartheta}$, $\hat{\phi}$ denote the temperature and the free boundary associated, respectively, with the nonlinear flux and initial temperatures g, ϑ_0 and $\hat{g}$, $\hat{\vartheta}_0$. We set

$$|g - \hat{g}|(z,t) = \sup_{\mu \leqslant \sigma \leqslant M} |g(z,t,\sigma) - \hat{g}(z,t,\sigma)| \quad \text{for a.e. } (z,t)\in \Sigma_N^T,$$

and we assume that the corresponding initial positions of the free boundaries are given by Lipschitz graphs, denoted by ϕ_0 and $\hat{\phi}_0$, respectively. Let $B = \sup_{\mu \leqslant \sigma \leqslant M} |b(\sigma)|$.

Theorem 2[R3]: Under the preceding notations, for each $t > 0$:

$$\int_I |\phi(t) - \hat{\phi}(t)| \leqslant \int_I |\phi_0 - \hat{\phi}_0| + \frac{B}{\lambda}\int_\Omega |\vartheta_0 - \hat{\vartheta}_0| + \frac{1}{\lambda}\int_0^t \int_0^L |g - \hat{g}| \equiv \delta(t).$$

In addition, for each $0 \leqslant \alpha < 1$ and $t > 0$, we have

$$\|\phi(t) - \hat{\phi}(t)\|_{C^{0,\alpha}(I)} \leqslant C_\alpha\,[\delta(t)]^{(1-\alpha)/2}$$

where the constant $C_\alpha > 0$ depends only of α, ℓ and L.

In order to discuss the asymptotic behaviour of the free boundary as $t \to \infty$, we recall some results on the time-independent problem, which we denote by the label ∞. For the boundary conditions (3),(4), with a time-independent flux $g_\infty = g_\infty(z,\vartheta)$ satisfying analogous assumptions, in [RY] it was shown the existence, uniqueness and continuous dependence of the weak solution $(\vartheta_\infty, \eta_\infty)\in[H^1(\Omega) \cap C^{0,\gamma}(\bar{\Omega})] \times L^\infty(\Omega)$, for some $0 < \gamma < 1$, of the continuous casting steady-state problem

$$\eta_\infty \in \beta(\vartheta_\infty) \text{ a.e. in } \Omega, \quad \vartheta_\infty = M \text{ on } \Gamma_0, \quad \vartheta_\infty = \mu \text{ on } \Gamma_L$$

146

$$\int_\Omega \nabla \vartheta_\infty . \nabla v - \nu \int_\Omega \eta_\infty \, \partial_z v = \int_{\Gamma_N} g_\infty (\vartheta_\infty) v, \quad \forall v \in H^1(\Omega) : v = 0 \text{ on } \Gamma_0 \cup \Gamma_L.$$

If, in addition, $\partial_z g_\infty \leqslant 0$, we have shown in [R2] that $\partial_z \vartheta_\infty \leqslant 0$, $\partial_x \vartheta_\infty \leqslant 0$, and, consequently, the steady-state free boundary $\Phi_\infty = \{\vartheta_\infty \geqslant 0\} \cap \{\vartheta_\infty \leqslant 0\}$ is the Lipschitz continuous curve

$$\Phi_\infty : \quad \zeta = \phi_\infty(\xi), \quad \text{with } |\phi'_\infty (\xi)| \leqslant 1 \quad \text{a.e. in } \xi \in I. \tag{13}$$

Now let $g(t) \to g_\infty$, as $t \to +\infty$, in the following sense

$$\int_0^\infty \int_0^L |g(t) - g_\infty| \, dz \, dt \leqslant C < +\infty \tag{14}$$

where we set, for $z \in [0,L]$ and $t \in]0,\infty[$,

$$|g(t) - g_\infty| (z) = \sup_{\mu \leqslant \sigma \leqslant M} |g(z,t,\sigma) - g_\infty (z,\sigma)|.$$

From Theorem 7 of [RY] we know that (14) implies the following asymptotic convergence as $t \to \infty$: $\vartheta(t) \to \vartheta_\infty$ in $L^p(\Omega)$, $\forall p < \infty$, and uniformly on compact subsets of Ω. Applying the continuous dependence arguments with $\hat{\eta}(t) \equiv \eta_\infty$ and $\hat{\phi}(t) \equiv \phi_\infty$, we can prove the following result.

Theorem 3[R3]: Under the preceding assumptions, namely (14), the evolutive free boundary $\phi(t)$ converges to the stationary free boundary ϕ_∞ in the following sense: $\phi(t) \to \phi_\infty$ in $C^{0,\alpha}(I)$, as $t \to +\infty$ ($0 \leqslant \alpha < 1$).

In [RY] it has also been shown that if, in addition, the initial data ϑ_0 and the evolutionary flux $g(x,t,\vartheta)$ are such that

$$\vartheta^\uparrow(t) \leqslant \vartheta(t) \leqslant \vartheta^\downarrow(t) \quad \text{for } t > 0, \tag{15}$$

where $\vartheta^\uparrow(t)$ and $\vartheta^\downarrow(t)$ are evolutive solutions, respectively, monotonously increasing and decreasing, as $t \to \infty$, towards the same steady-state solution ϑ_∞, then for each p, $1 \leqslant p < \infty$, there is a constant $C_p > 0$, such that

$$\|\vartheta(t) - \vartheta_\infty\|_{L^p(\Omega)} + \|\eta(t) - \eta_\infty\|_{L^p(\Omega)} \leqslant C_p \, t^{-1/p}.$$

This estimate for p=1 yields the following rate of asymptotic convergence.

Corollary: If, in addition to the assumptions of Theorem 3, the monotone frame (15) holds, in particular, for free boundaries monotone in time, we have the following estimates for t>0 and $0 \leqslant \alpha < 1$:

$$\|\phi(t) - \phi_\infty\|_{L^1_{(I)}} \leqslant \left(\frac{C_1}{\lambda t}\right) \quad \text{and} \quad \|\phi(t) - \phi_\infty\|_{C^{0,\alpha}_{(I)}} \leqslant C'_\alpha \, t^{(\alpha-1)/2}.$$

Remark 3: Sufficient conditions for the frame (15) to be verified are discussed in [RY]. For instance if $t \to g(t)$ is non-decreasing (resp. non-increasing) and if $\nu \, \partial_z \eta_o - \Delta \vartheta_o \leqslant 0$ (resp. $\geqslant 0$) then $t \to \vartheta(t)$ is non-decreasing (resp. non-increasing).

Remark 4: In [Z] close results have been established for a similar stationary problem in a infinite strip. Nevertheless our approach is different and provides a different study of the evolutionary free boundary. Our results on the free boundary for the two-phase continuous casting Stefan problem extend similar properties already known for the corresponding one-phase problem (see [R1], for references).

REFERENCES

[D]   **Danilyuk, I.I.** - On the Stefan problem, Russian Math. Surveys, **40**: 5 (1985), 157-223.

[M]   **Meirmanov, A.M.** - Stefan Problems (in Russian), Nauka, Novosibirsk, 1986.

[N]   **Nochetto, R.** - A class of non-degenerate two-phase Stefan problems in several space variables, Comm. P.D.E. **12** (1987), 21-45.

[R1]   **Rodrigues, J.F.** - "The Stefan problem revisited", in ISNM, Vol. **88**, Birkhäuser Basel (1989), 129-190.

[R2]   **Rodrigues, J.F.** - "On a steady-state two-phase Stefan problem with extraction" in ISNM, Vol. **95**, Birkhäuser Basel (1990), 229-240

[R3]   **Rodrigues, J.F.** - "On the free boundary of a two-phase Stefan problem with convection" CMAF Pre-print #13/90, Lisbon, 1990.

[RY]   **Rodrigues, J.F. & Yi, F.** - On a two-phase continuous casting Stefan problem with nonlinear flux, European J. Appl. Math. **1** (1990) 259-278.

[Z]   **Zaltzman, B.** - Multidimensional two-phase quasistationary Stefan problem in an unbounded region (Ph.D Thesis, Novosibirsk Univ., 1989).

José-Francisco Rodrigues, CMAF and University of Lisbon
Av. Prof. Gama Pinto, 2, 1699 Lisboa Codex, Portugal

C VUIK

The solution of a one-dimensional Stefan problem

1 Introduction

In this paper we prove existence and uniqueness of the solution of the following Stefan problem (compare [4]):

Problem 1
Given $T > 0$, the function C_0, the multifunction B and the functional G, find functions S and C such that:

$$\left.\begin{array}{rcll} \frac{\partial C(x,t)}{\partial t} - \frac{\partial^2 C(x,t)}{\partial x^2} & = & 0 \ , & x \in (-\infty, S(t)), \ t \in (0,T], \\ C(x,0) & = & C_0(x) \ , & x \in (-\infty, 0), \ S(0) = 0, \\ C(S(t),t) & = & 0 \ , & t \in [0,T], \end{array}\right\} \quad (1.1)$$

$$\left.\begin{array}{rll} G(S, f, t) \in B(S(t)) \ , & t \in [0,T], \\ \text{where } f(t) = \int\limits_{-\infty}^{\infty} [C_0(x) - C(x,t)]dx \ , & t \in [0,T], \end{array}\right\} \quad (1.2)$$

(for easy notation, we define $C(x,t) = 0$, $x \in (S(t), \infty)$, $t \in [0,T]$ and $C_0(x) = 0$, $x \in (0, \infty)$).

The most general existence and uniqueness results given in the literature are proved in [1] and [3]. The main differences between Problem 1 (and given results) and the results given in [1] are:

- Domain: In Problem 1 the diffusion equation is posed on an unbounded domain. In [1] the domain is bounded, but in [1; p.697] there are conjectures that results shown for a bounded domain can also be shown for an unbounded domain.

- Stefan condition: In [1] the Stefan condition is a differential equation. In Problem 1 the Stefan condition is a functional integral equation.

So Problem 1 can be used to describe a more general class of Stefan conditions than the class described in [1].

- Existence: In contrast with the existence results given in the literature we prove existence in a constructive way. So our proof suggests a numerical solution method. Another difference is that our conditions are such that the solution exists for every $T > 0$.

- Uniqueness: In the literature uniqueness of the solution is shown under the assumption that S is a differentiable function. We prove uniqueness under the assumption that S is a continuous function.

Our work was motivated by an etching technique which is used for the production of microelectronic devices [2], [5]. We use this application to illustrate our theory. In this case, the function C describes the concentration of the etching agent whereas $S(t)$ denotes the time-dependent position of the interface between solid and liquid. In most etching problems the vessel containing the liquid is very large with respect to the area wherein the etching agent shows a noticeable decrease. This is one of our reasons to pose the Stefan problem on an unbounded domain.

2 Existence and uniqueness

We start with a discussion of our Stefan condition (1.2). Integrating the well known Stefan condition

$$-\frac{\partial C}{\partial x}(S(t), t) = B_1 \frac{dS(t)}{dt} \ , \ t \in [0, T], \tag{2.1}$$

in the time direction and using the diffusion equation and the equation $\frac{\partial C}{\partial x}(-\infty, t) = 0$ gives

$$\int_{-\infty}^{\infty} [C_0(x) - C(x, t)]dx = B_1 S(t), \ t \in [0, T], \tag{2.2}$$

which is of the form (1.2) with $G(S, f, t) = f(t)$ and $B(x) = B_1 x$.

The Stefan condition (2.2) has the following features:

- Condition (2.2) is the mass balance in integral form. This means that the loss of etching agent is proportional to the loss of solid. The proportionality constant B_1 is given by the chemical properties of the etching agent and solid.

- The function S in (2.1) should be differentiable whereas in (2.2) it is sufficient that S be continuous. This is attractive from a physical point of view.

150

Obviously (1.2) is a generalization of (2.2). Since (1.2) does not look very natural we note that it was obtained by first proving some existence results for a Stefan problem using (2.2) and then trying to find the most general condition under which such kinds of proof could still be given.

In our existence proof we impose the following conditions:
Condition 2.3 ([4; p. 95, 96, 99])
We assume that the function C_0 should be a monotone decreasing Lipschitz continuous function with $C_0(0) = 0$ and $\lim_{x \to -\infty} C_0(x) = 1$. The multifunction B should be the inverse of a nondecreasing Lipschitz continuous function. The Lipschitz constant is denoted by $1/B_1$. The functional G should be Lipschitz continuous in its first and second argument with respect to the sup-norm. The Lipschitz constants are denoted by G_1 and G_2. Furthermore G should be a Lipschitz continuous function in its third argument (compare [4; p. 96, 2.3 v]).

We use following definitions:
Definition 2.4

(i) M is a function space (for details see [4; p. 98, 2.9])

(ii) The reduced problem is defined as follows: given a function S find a bounded solution C_S of the equations in (1.1) for this function S. The function $f_S : [0,T] \to \mathbb{R}$ is defined by

$$f_S(t) = \int_{-\infty}^{\infty} [C_0(x) - C_S(x,t)]dx.$$

(iii) The operator $T : M \to M$ is given by:

$$T(S)(t) = B^{-1}(G(S, f_S, t)), \quad t \in [0,T].$$

We have not imposed a sign condition on G. So the freezing of supercooled water can be described by Problem 1. In this application the speed of the interface $(S(t))$ between ice and water can be infinite. This is known as "blow up" of the speed of S. A large part of [4, Chapter 3] consists of the proof that there is no "blow up" and $T(S) \in M$ for all $S \in M$.

In the proof of our Main Theorem we use the following lemma:
Lemma 2.5
If $S_1(t) \le S_2(t)$, $t \in [0,T]$ and C_{S1}, C_{S2} exist then

$$\int_{-\infty}^{\infty} |C_{S2}(x,t) - C_{S1}(x,t)|dx \le \|S_1 - S_2\|_\infty, \ t \in [0,T].$$

Proof

Define $\delta = \|S_1 - S_2\|_\infty$. Using the maximum principle it can be shown that $C_{S2}(x + \delta, t) \leq C_{S1}(x, t) \leq C_{S2}(x, t)$. This implies

$$\int_{-\infty}^{\infty} |C_{S2}(x,t) - C_{S1}(x,t)|dx \leq \int_{-\infty}^{\infty} [C_{S2}(x,t) - C_{S2}(x + \delta, t)]dx \leq \|S_1 - S_2\|_\infty.$$

□

Main Theorem

If 2.3 holds and $G_1 + G_2 < B_1$ then Problem 1 has a unique solution.

Proof

Using 2.3, 2.4 and 2.5 we obtain

$$\|T(S_1) - T(S_2)\|_\infty \leq \frac{G_1 + G_2}{B_1}\|S_1 - S_2\|_\infty \text{ for } S_1, S_2 \in M.$$

Since $G_1 + G_2 < B_1$ this implies that the operator T is a contraction on the complete metric space $(M, \|.\|_\infty)$. According to the Banach fixed point theorem there is a unique function $\bar{S} \in M$ such that $T(\bar{S}) = \bar{S}$. This implies that $\{\bar{S}, C_{\bar{S}}\}$ is the solution of Problem 1. □

3 Applications of the Main Theorem

In this section we give two examples of Problem 1 and give an approximation of the solution. The numerically calculated iterates are computed with a numerical analogue of the relation $S_{i+1} = T(S_i)$.

Example 1

Define $B(x) = 2x$ and $G(S, f, t) = \begin{cases} -f(t) & , \quad t \in [0, 0.25] \\ -2f(0.25) + f(t) & , \quad t \in (0.25, 1] \end{cases}$.

A possible interpretation is: a solute precipitates until $t = 0.25$, thereafter the solute behaves itself as an etching agent. The numerically calculated iterates are given in Figure 1. Assuming that the third iterate is a good approximation of the solution it follows that the solid increases for $t \in [0, 0.25]$ and decreases for $t \in (0.25, 1]$. In $t = 0.25$ the function S is not differentiable.

Example 2

We consider the following "optimal control" problem. Suppose a layer of ice of $0°$is immersed in hot water. After a given time the ice should be melted. If the heat of the water is insufficient to achieve this result, then one uses a heat source with a given strength. To minimize energy costs the heat source is turned on as late as possible.

A mathematical model for this application is given by Problem 1 if B and G are defined as follows:

$$B(x) = 2x \text{ and } G(S, f, t) = f(t) + \int_0^t g(f, \tau)d\tau, \quad t \in [0, 1],$$

$$\text{where } g(f, t) = \begin{cases} 0 & , \quad t \in [0, 1 + (f(1) - 0.48)/0.75] \\ 0.75 & , \quad t \in (1 + (f(1) - 0.48)/0.75, \infty) \end{cases}.$$

In this model, C is the temperature of the water and $S(t)$ is the interface between ice and water. B describes the amount of heat needed to melt a certain amount of ice. The first term in G models the heat released from the water up to time t, whereras the second term models the heat obtained from the heat source. The quantities in the functional g can be interpreted as follows: the thickness of the ice layer is 0.24, the amount of heat needed to melt this layer is $0.48 = B(0.24)$, the ice should have melted at $t = 1$ and the strength of the heat source equals 0.75. We note that the heat source is not turned on $(g(f, t) = 0, t \in [0, 1])$ if the amount of heat $(f(1))$ released from the water at $t = 1$ is greater than the heat $(B(0.24))$ which is needed to melt the ice layer. On the other hand if $f(1) \leq B(0.24)$ then the heat source is turned on at $t_1 = 1 + (f(1) - 0.48)/0.75$. In this case suppose $\{\bar{S}, C_{\bar{S}}\}$ is the solution of Problem 1. It now follows from $G(\bar{S}, f, t) \in B(\bar{S}(t)), t \in [0, 1]$ and

$$G(\bar{S}, f, 1) = f(1) + \int_{t_1}^1 0.75 d\tau = f(1) - (f(1) - 0.48) = 0.48 \text{ that } \bar{S}(1) = \tfrac{1}{2}G(\bar{S}, f, 1) =$$

0.24. So the ice layer will have melted just in time.

The numerically calculated iterates are given in Figure 2. Assuming that the third iterate is a good approximation of the solution $\bar{S}$ we note that $\bar{S}(1) = 0.24$. Furthermore the calculation gives that the heat source should be turned on at $t_1 = 0.84$. As a result of this the slope of the graph of S shows a sudden increase at $t_1 = 0.84$.

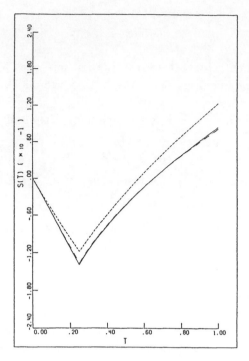

Figure 1   - - - iterate 1
        — — iterate 2
        —— iterate 3

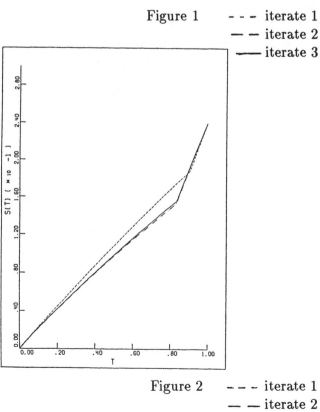

Figure 2   - - - iterate 1
        — — iterate 2
        —— iterate 3

Literature

1. Fasano, A., Primicerio, M.
   General free boundary problems for the heat equation I.
   J. Math. Anal. Appl. 57, pp. 694-723   (1977)

2. Notten, P.H.L.
   Electrochemical study of the etching of III-V semiconductors.
   Thesis, Eindhoven University of Technology   (1989)

3. Rubinstein, L.I.
   The Stefan problem.
   American Mathematical Society, Providence   (1971)

4. Vuik, C.
   The solutions of a class of Stefan problems.
   Thesis, State University Utrecht   (1988)

5. Vuik, C., Cuvelier, C.
   Numerical solution of an etching problem.
   J. Comp. Physics, 59, pp. 247-263   (1985)

C. Vuik
Delft University of Technology
Faculty of Technical Mathematics and Informatics
P.O. Box 5031, 2600 GA Delft, The Netherlands

Solidification II

T AIKI, N KENMOCHI AND J SHINODA

Periodic stability for Stefan problems with flux boundary conditions

1. Stefan problems with flux boundary conditions.

Let us consider Stefan problems with periodic condition in time in the enthalpy formulation:

(E) $u_t - \triangle\beta(u) = 0$ in $Q := J \times \Omega$,

(T) $u(t + T, x) = u(t, x)$ for $(t, x) \in Q$,

(B) some flux boundary condition on $\Sigma := J \times \Gamma$,

where Ω is a bounded domain in $\mathbf{R}^N (N \geq 1)$ with smooth boundary Γ, J is a time interval of the form (t_0, ∞), $-\infty \leq t_0 < \infty$, T is a given positive number, and $\beta : \mathbf{R} \to \mathbf{R}$ is a given Lipschitz continuous and nondecreasing function such that

$$\beta(0) = 0 \quad \text{and} \quad \liminf_{|r|\to\infty} \frac{\beta(r)}{r} > 0.$$

In this note we are interested in the asymptotic stability and order properties of the set of all solutions u to $\{(E), (T)\}$ (i.e. T-periodic solutions to (E)) with $J = \mathbf{R}$, provided the following type of boundary conditions (B_k), $k = 1, 2, 3$, is imposed on Σ as (B):

(B_1) (Usual non-homogeneous Neumann condition)

$$\frac{\partial\beta(u)}{\partial n} = h(t, x) \quad \text{on } \Sigma,$$

where h is a given function in $L^2_{loc}(\mathbf{R}; L^2(\Gamma))$ so that

$$\int_0^T \int_\Gamma h d\Gamma dt = 0, \quad h(t + T, x) = h(t, x) \quad \text{for } (t, x) \in \mathbf{R} \times \Gamma.$$

158

(B_2) (Nonlinear flux boundary condition)

$$-\frac{\partial \beta(u)}{\partial n} = \gamma(t, x, \beta(u)) \quad \text{on } \Sigma,$$

where $\gamma : \mathbf{R} \times \Gamma \times \mathbf{R} \to \mathbf{R}$ is a given function, satisfying the Caratheodory condition and the following $(\gamma 1) \sim (\gamma 4)$:

$(\gamma 1)$ $\gamma(t, x, r)$ is locally Lipschitz in $r \in \mathbf{R}$ (i.e. $\forall M > 0$, $\exists C_M > 0$ such that

$$|\gamma(t, x, r) - \gamma(t, x, r')| \leq C_M |r - r'|$$

for $\forall r, r' \in \mathbf{R}$ with $|r|, |r'| \leq M$);

$(\gamma 2)$ $\gamma(t, x, r)$ is nondecreasing in $r \in \mathbf{R}$, and $\gamma(\cdot, \cdot, r) \in L^2_{loc}(\mathbf{R}; L^2(\Gamma))$ for any $r \in \mathbf{R}$;

$(\gamma 3)$ there are constants M_1, M_2 with $M_1 \leq M_2$ such that

$$\gamma(t, x, \beta(M_1)) \leq 0, \quad \gamma(t, x, \beta(M_2)) \geq 0 \quad \text{a.e. on } \mathbf{R} \times \Gamma;$$

$(\gamma 4)$ $\gamma(t + T, x, r) = \gamma(t, x, r)$ for a.e. (t, x) and all r.

More generally, we consider sometimes a flux boundary condition of the form

(B_3) $\qquad -\dfrac{\partial \beta(u)}{\partial n} \in \gamma(t, x, \beta(u)) \quad \text{on } \Sigma,$

where $\gamma(t, x, r)$ is (multivalued) maximal monotone in $r \in \mathbf{R}$; for instance, $\gamma(t, x, r) = sign^+(r - g(t, x))$, g being a given function on Σ.

For the existence-uniqueness of solutions (in variational sense) to the Cauchy problems for $\{(E), (B_k)\}$, $k = 1, 2, 3$, we quote some results from [2,3,4,5,6]. By a solution of $\{(E), (B_1)\}$ on a compact time interval $I = [t_0, t_1]$ we mean a function $u : I \to L^2(\Omega)$ such that u is weakly continuous from I into $L^2(\Omega)$, $u \in W^{1,2}(I; (H^1(\Omega))')$ $((H^1(\Omega))'$ denotes the dual space of $H^1(\Omega))$, $\beta(u) \in L^2(I; H^1(\Omega))$ and

$$\int_I \langle u', \eta \rangle dt + \int_I \int_\Omega \nabla \beta(u) \cdot \nabla \eta \, dx \, dt = \int_I \int_\Omega h \eta \, d\Gamma dt \quad \text{for any } \eta \in L^2(I; H^1(\Omega)),$$

where $\langle \cdot, \cdot \rangle$ stands for the duality pairing between $(H^1(\Omega))'$ and $H^1(\Omega)$. For problem $\{(E), (B_2) \text{ (resp.}(B_3))\}$ its solution u is defined by replacing $h(t, x)$ by

$-\gamma(t, x, \beta(u))$ (resp. some function $h \in L^2(I; L^2(\Gamma))$ with $-h(t, x) \in \gamma(t, x, \beta(u))$ a.e. on Σ). Also, for a general time interval J and $k = 1, 2, 3$, we say that $u : J \to L^2(\Omega)$ is a solution of $\{(E), (B_k)\}$ on J, if it is a solution of $\{(E), (B_k)\}$ on every compact subinterval of J in the above sense. In the sequel, we use the following notations:

$$\mathcal{P}_k(T) = \{\text{all } T\text{-periodic solutions to } \{(E), (B_k)\} \text{ on } \mathbf{R}\},$$

$$\mathcal{P}_k(T; c) = \{u \in \mathcal{P}_k(T); \int_\Omega u(0, x)dx = c\}, \quad c \in \mathbf{R}.$$

2. Periodic stability for problem $\{(E), (B_1)\}$.

The first theorem is concerned with existence of T-periodic solutions to $\{(E), (B_1)\}$.

THEOREM 1. *For each $c \in \mathbf{R}$, $\mathcal{P}_1(T; c) \neq \emptyset$, that is, there is at least one T-periodic solution ω of $\{(E), (B_1)\}$ on $\mathbf{R}$ such that $\int_\Omega \omega(0, x)dx = c$.*

The next theorem is concerned with the structure of $\mathcal{P}_1(T; c)$.

THEOREM 2. *(a) If ω_1, $\omega_2 \in \mathcal{P}_1(T; c)$, then $\omega_1 - \omega_2$ is independent of t on $\mathbf{R} \times \Omega$,*

$$\int_\Omega (\omega_1(t, x) - \omega_2(t, x))dx = 0 \quad \text{for all } t \in \mathbf{R}$$

and

$$\beta(\omega_1) = \beta(\omega_2) \quad \text{on } \mathbf{R} \times \Omega.$$

(b) If $c_1 > c_2$ and $\omega_i \in \mathcal{P}_1(T; c_i)$, $i = 1, 2$, then

$$\beta(\omega_1) \geq \beta(\omega_2) \quad \text{on } \mathbf{R} \times \Omega.$$

Besides we see that for any $c \in \mathbf{R}$, $\mathcal{P}_1(T; c)$ is attractive (as $t \to \infty$) in the sense of the following theorem.

THEOREM 3. *For any solution u of $\{(E), (B_1)\}$ on $J = (t_0, \infty)$, with $\int_\Omega u(nT, x)dx = c$ for a large $n \in \mathbf{N}$, there exists $\omega \in \mathcal{P}_1(T; c)$ such that*

$$u(t) - \omega(t) \to 0 \quad \text{weakly in } L^2(\Omega) \text{ as } t \to \infty,$$

$$\beta(u(nT + \cdot)) \to \beta(\omega) \text{ in } L^2(0, T; H^1(\Omega)) \text{ as } n \to \infty.$$

Finally we have an interesting characterization of T-periodicity of solutions by means of global boundedness in $L^2(\Omega)$.

THEOREM 4. *Let u be a solution of $\{(E), (B_1)\}$ on $\mathbf{R}$. Then $u \in \mathcal{P}_1(T)$ if and only if $u \in L^\infty(\mathbf{R}; L^2(\Omega))$.*

The results mentioned in Theorem 1, Theorem 2(a) and Theorems 3,4 are due to [2], and Theorem 2(b) to [1]. These are quite applicable in studying similar properties of solutions to problem $\{(E), (B_2) \text{ or } (B_3)\}$.

3. Periodic stability for problem $\{(E), (B_2)\}$.

We recall the existence and uniqueness results on the Cauchy problem for $\{(E), (B_2)\}$ from [5,6], and taking advantage of L^∞-estimate for its solution, we can construct a T-periodic solution to $\{(E), (B_2)\}$.

THEOREM 5. $\mathcal{P}_2(T) \neq \emptyset$, and $\mathcal{P}_2(T) \subset L^\infty(\mathbf{R} \times \Omega)$.

For problem $\{(E), (B_2)\}$ a remarkable fact is stated in the following proposition.

PROPOSITION. *If ω_1, $\omega_2 \in \mathcal{P}_2(T)$, then*

$$\gamma(t, x, \beta(\omega_1)) = \gamma(t, x, \beta(\omega_2)) \quad \text{on } \mathbf{R} \times \Gamma,$$

$$\int_0^T \int_\Gamma \gamma(t, x, \beta(\omega_i)) d\Gamma dt = 0 \quad \text{for } i = 1, 2.$$

In the proof of the proposition the monotonicity of $\gamma(t, x, r)$ in $r \in \mathbf{R}$ is essential. By this proposition it is easily understood that all of facts about problem $\{(E), (B_1)\}$ with $h(t, x) = -\gamma(t, x, \beta(\omega))$ for $\omega \in \mathcal{P}_2(T)$ are applicable to $\{(E), (B_2)\}$, and we can prove the following theorem.

THEOREM 6. *(i) If ω_1, $\omega_2 \in \mathcal{P}_2(T)$ and*

$$\int_\Omega \omega_1(0, x) dx = \int_\Omega \omega_2(0, x) dx,$$

then $\omega_1 - \omega_2$ is independent of t on $\mathbf{R} \times \Omega$,

$$\int_\Omega \omega_1(t, x) dx = \int_\Omega \omega_2(t, x) dx \quad \text{for all } t \in \mathbf{R}$$

and

$$\beta(\omega_1) = \beta(\omega_2) \quad \text{on } \mathbf{R} \times \Omega.$$

(ii) If ω_1, $\omega_2 \in \mathcal{P}_2(T)$ with

$$\int_\Omega \omega_1(0, x) dx > \int_\Omega \omega_2(0, x) dx$$

then

$$\beta(\omega_1) \geq \beta(\omega_2) \quad \text{on } \mathbf{R} \times \Omega.$$

(iii) If ω_1, $\omega_2 \in \mathcal{P}_2(T)$ and

$$\int_\Omega \omega_1(0, x) dx > c > \int_\Omega \omega_2(0, x) dx,$$

then there is $\omega \in \mathcal{P}_2(T)$ such that

$$\int_\Omega \omega(0, x) dx = c.$$

(iv) Let u be any solution of $\{(E), (B_2)\}$ on (t_0, ∞). Then there is $\omega \in \mathcal{P}_2(T)$ such that

$$\beta(u(nT + \cdot)) \to \beta(\omega) \text{ in } L^2(0, T; H^1(\Omega)) \text{ as } n \to \infty.$$

Finally we mention two remarks (1), (2):

(1) Under some conditions on $\gamma(t, x, r)$, ensuring that the free boundary $\{(t, x);$ $\beta(u(t, x)) = 0\}$ never touches Σ, we can see that $\{(E), (B_3)\}$, with $\gamma(t, x, r) = sign^+(r - g(t, x))$, has a T-periodic solution and the set $\mathcal{P}_3(T)$ satisfies similar properties as $\mathcal{P}_2(T)$.

(2) For one dimensional two-phase Stefan problems we see under some regularity of β that the set $\mathcal{P}_k(T; c)$ $(k = 1, 2, 3)$ is a singleton for any $c \in \mathbf{R}$, and the T-periodic free boundaries associated to different constants c's do not meet each other.

References

1. T. Aiki, J. Shinoda and N. Kenmochi, *Periodic stability for some degenerate parabolic equations with nonlinear flux conditions*, Nonlinear Anal. T.M.A. **17** (1991), 885–902.

2. A. Haraux and N. Kenmochi, *Asymptotic behavior of solutions to some degenerate parabolic equations*, Funk. Ekvac. **34** (1991), 19–38.

3. N. Kenmochi and I. Pawlow, *The vanishing viscosity method in two-phase Stefan problems with nonlinear flux conditions*, Recent Topics in Nonlinear PDE IV, North-Holland, Amsterdam, (1988), 211–243.

4. M. Niezgodka and I. Pawlow, *A generalized Stefan problem in several space variables*, Appl. Math. Optim. **9** (1983), 193–224.

5. M. Niezgodka, I. Pawlow and A. Visintin, *Remarks on the paper by A. Visintin "Sur le problème de Stefan avec flux non linéaire"*, Boll. Un. Mat. Ital. 18C (1981), 87–88.

6. A. Visintin, *Sur le problème de Stefan avec flux non linéaire*, Boll. Un. Mat. Ital. 18C (1981), 63–86.

Toyohiko AIKI
Department of Mathematics,
Graduate School of Science and Technology,
Chiba University, Chiba-shi, 260 Japan

Nobuyuki KENMOCHI
Department of Mathematics, Faculty of Education,
Chiba University, Chiba-shi, 260 Japan

Junichi SHINODA
Department of Mathematics,
Graduate School of Science and Technology,
Chiba University, Chiba-shi, 260 Japan

G BELLETTINI[†], M PAOLINI[‡] AND C VERDI[#]

Convergence of discrete approximations to sets of prescribed mean curvature

An increasing interest has been recently brought up for attention by problems involving surface tension on interfaces or membranes and by their numerical approximations. The energy of such systems explicitly involves an unknown interface that makes very difficult the numerical treatment of the problem. Following the general setting proposed by E. De Giorgi [6], many functionals appearing in this context can be approximated by means of a sequence of regular nonconvex functionals. We observe that the lack of convexity creates further numerical problems, since we cannot expect the uniqueness of the solution. The relaxed functionals give rise to "thin" transition layers between two phases, and are closely related to several mathematical/physical models. We recall, among others, the phase field model [4], reaction-diffusion systems with fast reaction and slow diffusion [9,10], and Cahn Hilliard fluids [13]. The evolutionary case for such problems is of great importance and leads quite naturally to the study of the *motion by mean curvature* of surfaces [7,9,10].

Here we present a numerical approximation to a functional whose minimizers give rise to the so called interfaces with prescribed mean curvature. First this functional is regularized following an idea of L. Modica and S. Mortola [15], who proved convergence of the sequence of the relaxed functionals (see also [1,14,16]). Then it is discretized by means of piecewise linear finite elements with numerical quadrature [3], thus allowing the actual implementation on a computer. We have proved in [2] the convergence of the discrete minimizers to a solution of the continuous problem. Numerical experiments are in progress at the I.A.N. of C.N.R in Pavia and implementation details on the numerical algorithm, as well as several experiments, will appear in [3].

1. The prescribed curvature problem. Given an open bounded convex set $\Omega \subset \mathbf{R}^n$ ($n \geq 2$) with piecewise C^2 boundary, a function $\kappa \in L^\infty(\Omega)$, and $\mu \in [-1,1]$, we consider the minimum problem

$$(1.1) \quad \min_{A \subseteq \Omega} \tilde{\mathcal{F}}(A), \qquad \text{where} \qquad \tilde{\mathcal{F}}(A) := \mathcal{H}^{n-1}(\partial A \cap \Omega) + \mu \mathcal{H}^{n-1}(\partial A \cap \partial\Omega) - \int_A \kappa \, dx.$$

Here $\mathcal{H}^{n-1}$ denotes the $(n-1)$-dimensional Hausdorff measure [11]. The solution to this problem is a measurable set $A \subseteq \Omega$ whose boundary has mean curvature κ and contact angle $\theta = \arccos(\mu)$ at the intersection of ∂A with $\partial\Omega$ [12].

[†]International School for Advanced Studies SISSA/ISAS, 34014 Trieste, Italy.
[‡]Istituto di Analisi Numerica del CNR, 27100 Pavia, Italy.
[#]Dipartimento di Meccanica Strutturale, Università di Pavia, 27100 Pavia, Italy.

In order to introduce a rigorous formulation of problem (1.1) we need some notations. Let us denote by $BV(\Omega; \{-1, 1\})$ the space of the functions of bounded variation on Ω with values in $\{-1, 1\}$ and by $\mathrm{tr}(v)$ the trace on $\partial\Omega$ of the BV function v. Define the following energy functional on the space $D(\mathcal{F}) := BV(\Omega; \{-1, 1\})$

(C) $$\mathcal{F}(v) := \int_\Omega |Dv| + \int_{\partial\Omega} |\mathrm{tr}(v) + \mu| \, d\mathcal{H}^{n-1}(x) - \int_\Omega \kappa v \, dx,$$

where $\int_\Omega |Dv|$ denotes the total variation on Ω of the Radon measure Dv. It is well known [12] that $\mathcal{F}$ admits at least a minimizer u so that the set $A_u := \{x \in \Omega : u(x) = +1\}$ is a solution to problem (1.1). In addition, we have $\mathcal{F}(v) = 2\tilde{\mathcal{F}}(A_v) + (1 - \mu)\mathcal{H}^{n-1}(\partial\Omega) + \int_\Omega \kappa$, for all $v \in D(\mathcal{F})$. As a generalization, from now on we will consider μ to be a piecewise constant function $\mu \in BV(\partial\Omega; [-1, 1])$.

2. The relaxed problem. We first approximate $\mathcal{F}$ with a family $\{\mathcal{F}_\epsilon\}_{\epsilon > 0}$ of regular nonconvex functionals defined in $H^1(\Omega)$ which, in turn, will be discretized by finite elements; ϵ is the *relaxation* parameter and h is the *meshsize*. In order to define the relaxed functionals $\mathcal{F}_\epsilon$ we need some preparations. Let $\omega : [-1, 1] \to \mathbf{R}^+$ be defined by

(2.1) $$\omega(t) := 1 - t^2 \qquad [\text{or } \omega(t) := (1 - t^2)^2].$$

In addition, set

(2.2) $$\varphi(t) := \int_0^t \sqrt{\omega(s)} \, ds, \qquad \forall \, t \in [-1, 1],$$

(2.3) $$c_0 := \int_{-1}^1 \sqrt{\omega(t)} \, dt = 2\varphi(1) = -2\varphi(-1).$$

Since φ is an (odd) strictly increasing function, the following piecewise constant function $g \in BV(\partial\Omega; [-1, 1])$ is well defined

(2.4) $$g(x) := \varphi^{-1}\left(-\frac{c_0}{2}\mu(x)\right), \qquad \forall \, x \in \partial\Omega.$$

For any $\epsilon > 0$, let $g_\epsilon \in C^{0,1}(\partial\Omega; [-1, 1])$ approximate g in the sense that

(2.5) $$g_\epsilon(x) = g(x), \text{ if } \mathrm{dist}(x, S_g) \geq \epsilon, \qquad \mathrm{Lip}(g_\epsilon) \leq \frac{L_g}{\epsilon},$$

where S_g denotes the *jump* set of g and L_g is a constant independent of ϵ. It is obvious that $g_\epsilon \to g$ in $L^1(\partial\Omega)$, as $\epsilon \to 0$.

We are now in a position to define the relaxed functionals as

(R) $$\mathcal{F}_\epsilon(v) := \int_\Omega \left[\epsilon|\nabla v|^2 + \frac{1}{\epsilon}\omega(v) - c_0\kappa v\right] dx, \quad \text{if } v \in D(\mathcal{F}_\epsilon),$$

165

where $D(\mathcal{F}_\epsilon) := \{v \in H^1(\Omega; [-1,1]) : v = g_\epsilon \text{ on } \partial\Omega\}$. The existence of a solution u_ϵ to the minimum problem for $\mathcal{F}_\epsilon$ can be proven by direct methods. We have that $\mathcal{F}_\epsilon \xrightarrow{\Gamma} c_0 \mathcal{F}$, as $\epsilon \to 0$ [1,16] (we refer to [8] for basic issues about Γ-convergence). The nonconvexity of $\mathcal{F}_\epsilon$ allows, in general, the presence of more than one absolute or relative minima u_ϵ satisfying the following variational inequality

$$(2.6) \qquad 2\epsilon\langle\nabla u_\epsilon, \nabla v - \nabla u_\epsilon\rangle + \frac{1}{\epsilon}\langle F(u_\epsilon), v - u_\epsilon\rangle - c_0\langle\kappa, v - u_\epsilon\rangle \geq 0, \qquad \forall v \in D(\mathcal{F}_\epsilon),$$

where $\langle u, v\rangle := \int_\Omega uv \, dx$ and $F(s) := \omega'(s)$, for all $s \in [-1, 1]$.

The choice $\omega(t) := (1 - t^2)^2$ has been considered in papers where more regularity is required [9]. Moreover, with this choice of ω, now defined on the whole $\mathbf{R}$, v is allowed to assume values outside $[-1, 1]$. Hence we can define $D(\mathcal{F}_\epsilon) := \{v \in H^1(\Omega) : v = g_\epsilon \text{ on } \partial\Omega\}$ and thereafter the variational inequality (2.6) is replaced by the Euler-Lagrange equation $-2\epsilon\Delta u_\epsilon + \frac{1}{\epsilon}F(u_\epsilon) - c_0\kappa = 0$.

On the other hand the choice $\omega(t) = 1 - t^2$ is useful because transitions layers with thickness $C\epsilon$ are allowed. This fact has been proved in one dimension [2].

3. The discretized problem. At this stage, we can introduce a discretization of the relaxed functional $\mathcal{F}_\epsilon$ by piecewise linear finite elements. Let us denote by $\{S_h\}_{h>0}$ a regular and weakly acute family of partitions of Ω into simplices having diameter bounded by h [5, p. 132]. For the sake of simplicity we shall assume that Ω is a polyhedron, so that $\bar{\Omega} \equiv \Omega_h := \cup_{S \in S_h} S$, for all $h > 0$. Let $\mathbf{V}_h \subset H^1(\Omega)$ indicate the usual piecewise linear finite element space over S_h and $\Pi_h : C^0(\bar{\Omega}) \to \mathbf{V}_h$ the Lagrange interpolation operator. Let $\{a_j\}_{j=1}^J$ denote the nodes of S_h and $\{\chi_j\}_{j=1}^J$ the canonical basis of $\mathbf{V}_h$; we assume that $\{a_j\}_{j=1}^{J_0}$ are the internal nodes.

Finally, for any $\epsilon > 0$, let $\kappa_\epsilon \in C^{0,1}(\bar{\Omega})$ approximate the *curvature* function κ so that

$$(3.1) \qquad \|\kappa_\epsilon\|_{L^\infty(\Omega)} \leq C, \qquad \mathrm{Lip}(\kappa_\epsilon) \leq \frac{L_\kappa}{\epsilon}, \qquad \kappa_\epsilon \underset{\epsilon \to 0}{\to} \kappa \text{ in } L^1(\Omega),$$

where C and L_κ are constants independent of ϵ.

We are now in a position to introduce the discrete functionals

$$(D) \qquad \mathcal{F}_{\epsilon,h}(v) := \int_\Omega \left[\epsilon|\nabla v|^2 + \frac{1}{\epsilon}\Pi_h\omega(v) - c_0\Pi_h(\kappa_\epsilon v)\right] dx, \quad \text{if } v \in D(\mathcal{F}_{\epsilon,h}),$$

where $D(\mathcal{F}_{\epsilon,h}) := \{v \in \mathbf{V}_h : |v| \leq 1 \text{ in } \Omega, \ v = \Pi_h g_\epsilon \text{ on } \partial\Omega\}$. The existence of a solution $u_{\epsilon,h}$ to the minimum problem for $\mathcal{F}_{\epsilon,h}$ is trivial. The integrals in (D) can be evaluated via the vertex quadrature rule, which is exact for piecewise linear functions, thus allowing the actual implementation of the discrete minimum problem on a computer.

166

A (relative) minimum $u_{\epsilon,h}$ for the functional $\mathcal{F}_{\epsilon,h}$ must satisfy the variational inequality
(3.2)
$$2\epsilon\langle\nabla u_{\epsilon,h},\nabla v-\nabla u_{\epsilon,h}\rangle+\frac{1}{\epsilon}\langle F(u_{\epsilon,h}),v-u_{\epsilon,h}\rangle_h-c_0\langle\kappa_\epsilon,v-u_{\epsilon,h}\rangle_h\geq 0,\qquad\forall\,v\in D(\mathcal{F}_{\epsilon,h}),$$

where $\langle u,v\rangle_h:=\int_\Omega\Pi_h(uv)dx$. Denoting by

$$\begin{aligned}M&:=\{\langle\chi_j,\chi_k\rangle_h\}_{j,k=1}^J,&&\text{(diagonalized) mass matrix,}\\(3.3)\qquad A&:=\{\langle\nabla\chi_j,\nabla\chi_k\rangle\}_{j,k=1}^J,&&\text{stiffness matrix,}\\D&:=\text{diagonal of }A,&&B:=I-D^{-1}A,\end{aligned}$$

the variational inequality (3.2) can then be written in matrix form as follows

$$(3.4)\qquad\begin{cases}[2\epsilon AU+\frac{1}{\epsilon}MF(U)-c_0MK]_j\in\psi(u_j),&u_j\in[-1,1],&j=1,...,J_0,\\u_j=g_\epsilon(a_j),&&j=J_0+1,...,J.\end{cases}$$

Here $U=(u_1,...,u_J)$ is the vector representing the unknown solution $u_{\epsilon,h}$ ($u_j:=u_{\epsilon,h}(a_j)$, $j=1,...,J$), $K:=(\kappa_\epsilon(a_1),...,\kappa_\epsilon(a_J))^T$, and $-\psi$ is the inverse of the *sgn* graph. In other words, $z\in\psi(u_j)$ means $z=0$ if $u_j\in(-1,1)$, $z\leq 0$ if $u_j=1$, $z\geq 0$ if $u_j=-1$.

A simple minded method for solving the nonlinear system (3.4) is the following Jacobi iterative scheme

$$(3.5)\qquad u_j^{k+1}:=\Pi_{[-1,1]}\left(BU^k-\frac{1}{2\epsilon^2}D^{-1}MF(U^k)+\frac{c_0}{2\epsilon}D^{-1}MK\right)_j,\qquad j=1,...,J_0,$$

where $\Pi_{[-1,1]}(s):=\min(1,\max(-1,s))$. A more effective iterative method can be easily obtained by using an incomplete Cholesky factorization of the matrix A as a preconditioner.

Standard compactness techniques [3] give the convergence of minimizers of (D) to a minimum of (R), as $h\to 0$.

4. Relationship between ϵ and h. We have seen that (D)→(R) as $h\to 0$ and that (R)→(C) as $\epsilon\to 0$. It is important then to understand what happens if we let ϵ and h go to zero simultaneously. In other words, how should ϵ and h be related in order to have (D)→(C) as both go to zero? The answer to this question is given by the following Theorem, which is fully proven in [2].

THEOREM 4.1. *Let $h=o(\epsilon)$. Then the sequence $\{\mathcal{F}_{\epsilon,h}\}_{\epsilon,h}$ Γ-converges to $c_0\mathcal{F}$, as $\epsilon\to 0$. Moreover, any family $\{u_{\epsilon,h}\}_{\epsilon,h}$ of absolute minimizers of $\{\mathcal{F}_{\epsilon,h}\}_{\epsilon,h}$ is relatively compact in $L^1(\Omega)$, and any of its limit points minimizes $\mathcal{F}$.*

5. Numerical experiments. We now present some numerical results for two different examples. In the first example we take $\Omega:=(0,4)\times(0,4)$, $\mu=0$ (normal contact) and $\kappa(x):=-m|x|+2m+\frac{1}{2}$, where $|x|$ is the euclidean norm of $x\in\Omega$ and m is a parameter

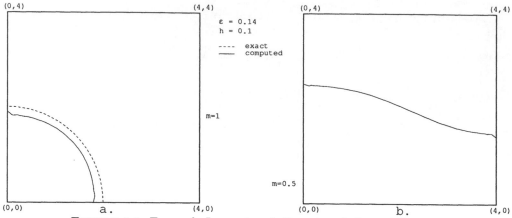

FIGURE 5.1. Example I: exact and discrete relative minima.

giving the slope of κ. The empty set is always a relative minimum and, if $m > \frac{1}{4}$, another radial-symmetric relative minimum is the portion of the circle centered at the origin with radius $r = 2$ lying in Ω. This is the absolute minimum if $m > \frac{3}{4}$, at least among the radial-symmetric subsets of Ω. In Figure 5.1a the result of the computation with $m = 1$ is shown. However also non radial-symmetric solutions exist, as shown in Figure 5.1b $(m = 0.5)$.

In the second example we take $\Omega := (0,1) \times (0,1)$, $\mu = 1$ (tangential contact) and $\kappa(x) = \lambda$, where λ is a positive real number. The empty set is again a relative minimum, whereas the absolute minimum (if λ is sufficiently high) is Ω with the corners smoothed in circular arcs of radius $\frac{1}{\lambda}$. Figure 5.2a $(\lambda = 4)$ shows that actually the minimizing set A presents no contact with $\partial\Omega$. This is due to the fact that the relaxed solution has a transition layer of width $C\epsilon$ with values ranging from -1 to 1 in a strip along $\partial A \cap \partial\Omega$. Figure 5.2b $(\lambda = 8)$ shows the discrete and exact solutions in a nonconvex domain Ω.

We note, as a final remark, that the number of Jacobi iterations necessary to reach convergence of the scheme is quite high (e.g., the result shown in Figure 5.2a has been obtained after 2216 iterations). This behaviour can be explained by noting that the Jacobi iterative scheme (3.5) can be viewed as an explicit Euler time discretization of the evolution equation

$$(5.1) \qquad u_t = \Delta u - \frac{1}{2\epsilon^2} F(u) + \frac{c_0}{2\epsilon} \kappa$$

with a time step $\tau = Ch^2$. This equivalence can be easily shown when the mesh S_h is made of equilateral triangles of side h. Equation (5.1) describes the *motion by mean curvature* plus the extra *prescribed curvature* term $\frac{c_0}{2\epsilon}\kappa$. Hence the interface is expected to move with speed proportional to its mean curvature (diminished by κ) and to approach the limit position within a given threshold distance in finite time. We can then expect that a number of time steps of order $O(h^{-2})$ is required to reach convergence for both (5.1) and (3.5).

168

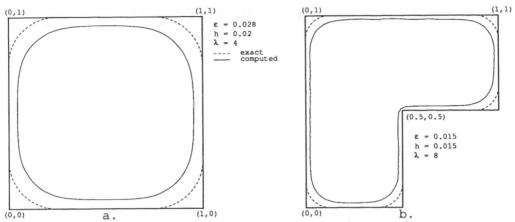

FIGURE 5.2. Example II: exact and discrete relative minima.

REFERENCES

[1]   S. BALDO AND G. BELLETTINI, Γ-convergence and numerical analysis: an application to the minimal partitions problem, Ricerche di Matematica (to appear).

[2]   G. BELLETTINI, M. PAOLINI, AND C. VERDI, Γ-convergence of discrete approximations to interfaces with prescribed mean curvature, Atti Accad. Naz. Lincei Rend. Cl. Sci. Fis. Mat. Natur. (9) (to appear).

[3]   G. BELLETTINI, M. PAOLINI, AND C. VERDI, Numerical minimization of geometrical type problems related to calculus of variations, Calcolo (to appear).

[4]   G. CAGINALP, An Analysis of a Phase Field Model of a Free Boundary, Arch. Rational Mech. Anal., 92 (1986), pp. 205–245.

[5]   P.G. CIARLET, The Finite Element Method for Elliptic Problems, North-Holland, Amsterdam, 1978.

[6]   E. DE GIORGI, Free discontinuity problems in calculus of variations, in Proc. Meeting in honour of J.L. Lions, North-Holland, Amsterdam, 1988, to appear.

[7]   E. DE GIORGI, Some conjectures on flow by mean curvature. Conference held at the University of Telaviv.

[8]   E. DE GIORGI AND T. FRANZONI, Su un tipo di convergenza variazionale, Atti Accad. Naz. Lincei Rend. Cl. Sci. Fis. Mat. Natur. (8), 58 (1975), pp. 842–850.

[9]   P. DE MOTTONI AND M. SCHATZMAN, Geometrical evolution of developped interfaces, publ. Equipe d'Analyse Numérique Lyon Saint-Etienne, 87 (1989), pp. 1–68.

[10] L.C. EVANS AND J. SPRUCK, Motion of level sets by mean curvature. Preprint.

[11] H. FEDERER, Geometric Measure Theory, Springer-Verlag, Berlin, 1968.

[12] R. FINN, Equilibrium Capillary Surfaces, Springer-Verlag, Berlin, 1986.

[13] M.E. GURTIN, On phase transitions with bulk, interfacial, and boundary energy, Arch. Rational Mech. Anal., 96 (1986), pp. 243–264.

[14] L. MODICA, The gradient theory of phase transitions and minimal interface criterion, Arch. Rational Mech. Anal., 98 (1987), pp. 123–142.

[15] L. MODICA AND S. MORTOLA, Un esempio di Γ-convergenza, Boll. Un. Mat. Ital. (5), 14 B (1977), pp. 285–299.

[16] N.C. OWEN, J. RUBINSTEIN, AND P. STERNBERG, Minimizer and gradient flows for singularly perturbed bi-stable potentials with a Dirichlet condition. Preprint.

J D FEHRIBACH

A continuation method for a self-similar moving boundary problem

Consider the system

$$(a) \quad p_t = [D_p p_x]_x + v p_x, \quad x \in (0, \infty), \quad t > 0,$$

$$(b) \quad n_t = [D_n n_x]_x + v n_x, \quad x \in (0, \infty), \quad t > 0,$$

$$(c) \quad v(p_s - p) = D_p p_x, \quad x = 0, \quad t > 0,$$

$$(d) \quad v(n_s - n) = D_n n_x, \quad x = 0, \quad t > 0, \qquad (1)$$

$$(e) \quad f(p, n) = 0, \qquad x = 0, \quad t > 0,$$

$$(f) \quad p = p_\infty, \quad x \in (0, \infty), \quad t = 0,$$

$$(g) \quad n = n_\infty, \quad x \in (0, \infty), \quad t = 0.$$

This system is a model for protein crystal growth from a solution containing protein and a precipitant dissolved in a solvent. It has been previously studied by Fehribach & Rosenberger (1989). The notation is as follows: x measures the distance from the solid-liquid interface, p and n represent the concentrations of two diffusing species, p_s and n_s are the constant, uniform concentrations in the solid, p_∞ and n_∞ are uniform initial concentrations, v is the interface velocity, and $f = 0$ represents a curve and guarantees that the solid and liquid phases are in equilibrium at the interface. The diffusivities are assumed to depend on the concentrations (cf., Chang & Myerson (1985)), viz. $D_p \equiv D_p(p, n)$ and $D_n \equiv D_n(p, n)$, and to be sufficiently smooth and positive (though possibly small) for all values of the concentrations. The triple (p, n, v) is called a solution if it satisfies (1).

Presented here is a continuation method for computing similarity solutions for self-similar moving boundary problems such as (1). Briefly, the approach is to consider a nontrivial self-similar semi-infinite or infinite moving boundary problem, insert a continuation parameter, make an appropriate restriction of the problem to a finite interval, and finally compute a homotopy of solutions leading from a known solution when the continuation parameter is zero to the desired solution when the continuation parameter is, say, one. The restriction of the problem to a finite interval is accomplished by adapting the work of Friedman & Doedel (1991). The homotopy of solutions is computed using the numerical package AUTO [Doedel & Kernevez (1986)].

Note that (1) is invariant under a well-known similarity transformation: $(x,t) \rightarrow (\beta x, \beta^2 t)$. Hence define $\xi \equiv x/2\sqrt{t}$, and write (1) as a first order system with the continuation parameter α inserted in the diffusivities:

$$
\begin{aligned}
(a) \quad & u_1' = u_3, \\
(b) \quad & u_2' = u_4, \\
(c) \quad & u_3' = -\frac{2(\xi + \lambda)u_3 + \alpha \nabla D_p \cdot (u_3, u_4)u_3}{D_p(\alpha u_1, \alpha u_2)}, \\
(d) \quad & u_4' = -\frac{2(\xi + \lambda)u_4 + \alpha \nabla D_n \cdot (u_3, u_4)u_4}{D_n(\alpha u_1, \alpha u_2)},
\end{aligned}
\tag{2}
$$

$$
\begin{aligned}
(e) \quad & 2\lambda(p_s - u_1) = D_p(\alpha u_1, \alpha u_2)u_3, \quad \xi = 0, \\
(f) \quad & 2\lambda(n_s - u_2) = D_n(\alpha u_1, \alpha u_2)u_4, \quad \xi = 0, \\
(g) \quad & f(u_1, u_2) = 0, \quad \xi = 0, \\
(h) \quad & u(\infty) = u_\infty \equiv (p_\infty, n_\infty, 0, 0)
\end{aligned}
$$

where $u \equiv (p, n, p_\xi, n_\xi)$ and $\lambda \equiv v\sqrt{t}$. System (2) is of the form $u' = F(\xi, u, \lambda, \alpha) \equiv F_0(u, \alpha) + (\xi + \lambda)F_1(u, \alpha)$, and as $\xi \rightarrow \infty$, one can show $F(u, \lambda, \alpha, \xi) \sim (\xi + \lambda)F_1(u, \alpha)$. By direct computation one then finds that the matrix $F_{1u}(u_\infty, \alpha)$ is diagonal with eigenvalues $(\mu_1, \mu_2, \mu_3, \mu_4) = (-2/D_p(\alpha p_\infty, \alpha n_\infty), -2/D_n(\alpha p_\infty, \alpha n_\infty), 0, 0)$ and hence that the stable manifold at $\xi = \infty$ is two dimensional (stable manifold theorem using $\zeta \equiv (\xi + \lambda)^2/2$). Therefore to restrict to a finite interval, adjoin to (2) the equations

$$
\begin{aligned}
(a) \quad & F_{1u}(u_\infty, \alpha)v_i = \mu_i v_i, \quad i \in \{1, 2\}, \\
(b) \quad & u(\xi_R) = u_\infty + c_1 v_1 + c_2 v_2.
\end{aligned}
\tag{3}
$$

where ξ_R is the right computational endpoint, v_1 and v_2 are the corresponding eigenvectors, and c_1 and c_2 are constants. Eqn. (3b) is then backsubstituted into (2) and (3a) to eliminate u_∞. This leaves four equations with seven side conditions. At least formally then, one would expect three scalar parameters which are to be computed in the continuation process, viz. λ, c_1, c_2. The computational right endpoint ξ_R is fixed throughout the continuation.

The local justification (α sufficiently small) for the above process parallels that presented by Friedman & Doedel (1991) for heteroclinic orbits. The goal is to show that in the correctly weighted Banach spaces, an appropriate operator is Fredholm, and an implicit function theorem of Descloux & Rappaz can be applied to establish existence, uniqueness and error estimates for the approximate problem. The three lemmas which follow establish the hypotheses needed to apply this implicit function theorem. Their proofs are given in Fehribach (1992).

Let μ be a negative real such that for all α, $\mu_i < \mu < 0$. For $\xi > 0$, define $\gamma_\mu(\xi) \equiv e^{-\mu\xi^2/2}$. Also define the norms $\|u\|_{\mu,0} \equiv \max_{1 \le i \le 4} \|\gamma_\mu u_i\|_{L_\infty}$, and

$$\|u\|_{\mu,j} \equiv \|u\|_{\mu,0} + \|du/dt\|_{\mu,0} + ... + \|d^j u/dt^j\|_{\mu,0}, \ j \ge 0.$$

Finally let $B_{\mu,j}$ be the Banach space of vector valued functions on $(0,\xi)$ with finite $\|\cdot\|_{\mu,j}$ norm. Because of the asymptotic behavior of (2), one is led to look for solutions of the form $u(\xi) = w(\xi) + u_\infty$ where $w \in B_{\mu,1}$. In this regard, consider the operator $\Phi : B_{\mu,1} \times \mathbf{R}^2 \to B_{\mu,0}$ defined by

$$\Phi(\cdot, w, \lambda, \alpha) \equiv w' - F(\cdot, w + u_\infty, \lambda, \alpha).$$

Let $(w_0, \lambda^0, 0) \in B_{\mu,1} \times \mathbf{R}^2$ be a solution of $\Phi = 0$ and define the linearized operator $\Phi_w^0 : B_{\mu,1} \to B_{\mu,0}$ by

$$\Phi_w^0 \equiv \Phi_w(\cdot, w_0 + u_\infty, \lambda^0, 0) = \frac{d}{d\xi} - A(\cdot),$$

where

$$A(\xi) \equiv F_u(\xi, w_0(\xi) + u_\infty, \lambda^0, 0).$$

Lemma 1: *The operator $\Phi_w^0 : B_{\mu,1} \to B_{\mu,0}$ is a Fredholm operator with*

$$\dim \mathcal{N}(\Phi_w^0) = \mathrm{codim}\mathcal{R}(\Phi_w^0) = 2.$$

That $\mathrm{codim}\,\mathcal{R}(\Phi_w^0) = 2$ is a problem for the argument below. However, if one considers the original system $\Phi = 0$, one sees that $\mathcal{R}(\Phi)$ is also annihilated by the same two dimensional subspace of constants. Therefore

172

define $R_{\mu,0} \equiv \mathcal{R}(\Phi_w^0)$ with norm as in $B_{\mu,0}$. Then $\Phi_w^0 : B_{\mu,1} \to R_{\mu,0}$ is Fredholm with index two.

For there to be a unique solution, the interface conditions (1.1c-e) must be included. Define the operator $\Psi : B_{\mu,1} \times \mathbf{R}^2 \to R_{\mu,0} \times \mathbf{R}^3$ as

$$\Psi(\cdot, w, \lambda, \alpha) \equiv (\Phi(\cdot, w, \lambda, \alpha), I(w, \lambda, \alpha))$$

where

$$I(w, \lambda, \alpha) \equiv \begin{cases} 2\lambda(p_s - u_1) - D_p(\alpha u_1, \alpha u_2)u_3 \\ 2\lambda(n_s - u_2) - D_n(\alpha u_1, \alpha u_2)u_4 \;. \\ \qquad f(u_1, u_2) \end{cases}$$

The following lemma extends the results of Lemma 1 to the operator Ψ.

Lemma 2: *Define the operator* $\Psi'(\cdot, w_0, \lambda^0, 0) : B_{\mu,1} \times \mathbf{R} \to R_{\mu,0} \times \mathbf{R}^3$ *as*

$$\Psi'(\cdot, w_0, \lambda^0, 0) \equiv \begin{bmatrix} \Phi_w^0 & \Phi_\lambda^0 \\ I_w^0 & I_\lambda^0 \end{bmatrix}.$$

This operator is Fredholm (of index 0) with $\mathcal{R}(\Psi') = R_{\mu,0} \times \mathbf{R}^3$.

Now define an approximate operator Ψ_T with the appropriate asymptotic behavior as $\xi \to \infty$:

$$\Psi_T(\cdot, w, \lambda, \alpha) \equiv (\Phi_T(\cdot, w, \lambda, \alpha), I(w, \lambda, \alpha))$$

where

$$\Phi_T(\xi, w, \lambda, \alpha) \equiv \begin{cases} \Phi(\xi, w, \lambda, \alpha) & \xi \leq \xi_R \\ w' - (\xi + \lambda)F_{1u}(u_\infty, \lambda, \alpha)w & \xi > \xi_R \end{cases}.$$

The next Lemma establishes a pointwise convergence condition for Ψ and Ψ_T and a convergence condition on the derivatives at the initial solution.

Lemma 3: *For the operators defined above,*

$$\|\Psi(\cdot, w, \lambda, \alpha) - \Psi_T(\cdot, w, \lambda, \alpha)\| \leq C\|w\|_{\mu,0}^2 e^{\mu \frac{\xi_R^2}{2}} \to 0 \quad as \quad \xi_R \to \infty$$

$$\|\Psi'(\cdot, w_0, \lambda^0, 0) - \Psi_T'(\cdot, w_0, \lambda^0, 0)\| \leq C e^{\mu \frac{\xi_R^2}{2}} \to 0, \quad as \quad \xi_R \to \infty$$

where $\mu < 0$ *and C does not depend on* λ *or* ξ_R.

The above lemmas are sufficient to establish the hypotheses of a simplified version of the implicit function theorem (Theorem 3.1, p.326) from Descloux and Rappaz (1982) (cf. also Theorem 1, Friedman and Doedel (1991)).

Theorem 1: *There exists positive constants ξ_R^0, α_0, ϵ, C and two unique C^1 maps $x(\alpha) \equiv (w(\alpha), \lambda(\alpha))$ and $x_T(\alpha) \equiv (w_T(\alpha), \lambda_T(\alpha))$ satisfying for $|\alpha| < \alpha_0$ and $\xi_R > \xi_R^0$ the following conditions:*

$$\Psi(\cdot, x(\alpha), \alpha) = 0, \quad \|x(\alpha) - x_0\|_{B_{\mu,1} \times \mathbf{R}} < \epsilon,$$

$$\Psi_T(\cdot, x_T(\alpha), \alpha) = 0, \quad \|x_T(\alpha) - x_0\|_{B_{\mu,1} \times \mathbf{R}} < \epsilon,$$

and

$$|\lambda(\alpha) - \lambda_T(\alpha)| + \|w(\alpha) - w_T(\alpha)\|_{\mu,1} \le C\|w(\alpha)\|_{\mu,1} e^{\mu \frac{\xi_R^2}{2}}.$$

Now consider a simple example: let $D_p = D_p^0(1 - 3\alpha p)$ and $D_n = D_n^0$. Notice that for this D_p, α is proportional to $\left|\frac{\partial D_p}{\partial p}\right|$. Let $p_\infty = 1.6 \times 10^{-3}$, $n_\infty = 2.0 \times 10^{-2}$, $p_s = 0.7$, $n_s = 0.3$ (all in mole fractions). Also let $D_p^0 = 1.0 \times 10^{-7}$, $D_n^0 = 1.0 \times 10^{-5}$ (cm^2/sec). Finally let $f(p, n) = 1000p + 10n - 1$. These parameter values along with this linear solubility function are take from Fehribach & Rosenberger (1989).

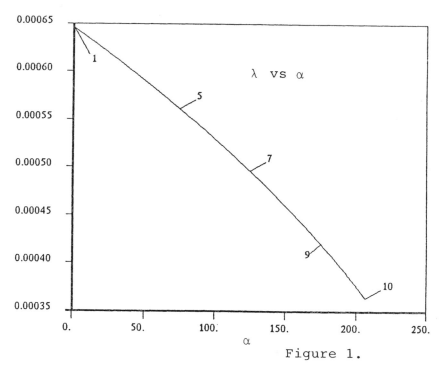

Figure 1.

174

The continuation computation for this system is presented in Figures 1 and 2. As would be expected, the continuation can be performed until the diffusivities approach zero (the scaling on α is a matter of convenience; here $D_p \simeq 0$ for $\alpha \simeq 200$ where the equation for protein diffusion is essentially a porous media equation). Figure 1 is a plot of the growth parameter λ vs α. Figures 2 shows the profiles p and p_ξ for various labeled points on the λ-α curve. In both figures, distances have been scaled by $\sqrt{D_p^0}$. Notice that as α increases, the profile of p_ξ becomes linear for small ξ with a shock at $p = p_\infty$ (i.e., a corner shock in the protein concentration at $p = p_\infty$). Thus in addition to the original phase interface at $x = 0$, a second free boundary has developed. For $\alpha \gg 1$ (but $\alpha \lesssim 1/3p_\infty$), an asymptotic argument shows that for the given D_p there is a solution with p_ξ approximately linear. This behavior is consistent with Figure 2 and quite different from what would occur if D_p were small but independent of concentration. In that case the profile of p would always be an error function and $\lambda = O(\sqrt{D_p})$.

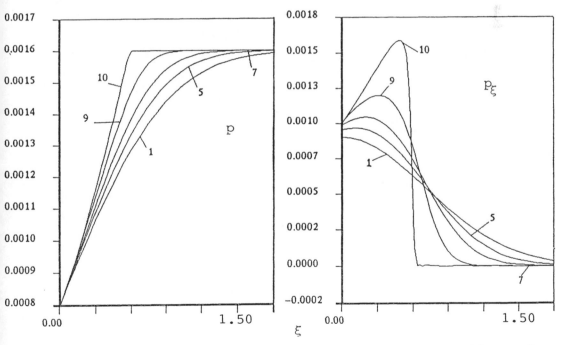

Figure 2.

Acknowledgments

The author wishes to thank Mark Friedman for a number of helpful discussions and wishes to acknowledge the support of the Center for Microgravity and Materials Research of the University of Alabama in Huntsville.

References

Chang, Y.C. & A.S. Myerson, The diffusivity of potassium chloride and sodium chloride in concentrated, saturated, and supersaturated aqueous solutions, *AlChE J.* **31** (1985) 890–894.

Descloux, J. & J. Rappaz, Approximation of solution branches of nonlinear equations, *R.A.I.R.O. Analyse numérique/Num. Anal.* **16** (1982) 319–349.

Doedel, E.J., & J.P. Kernévez, AUTO: Software for continuation and bifurcation problems in ordinary differential equations, Applied Mathematics Report, California Institute of Technology (1986).

Fehribach, J.D., Analysis and application of a continuation method for a self-similar coupled Stefan system, to appear in *Q. Appl. Math.* (1992).

Fehribach, J.D., & F. Rosenberger, Analysis of models for two solution crystal growth problems, *J. Crystal Growth* **94** (1989) 6–14.

Friedman, M.J., & E.J. Doedel, Numerical computation and continuation of invariant manifolds connecting fixed points, *SIAM J. Numer. Anal.* **28** (1991) 789–808.

Department of Mathematical Sciences
University of Alabama in Huntsville
Huntsville, AL 35899

R H NOCHETTO[†], M PAOLINI[‡] AND C VERDI[#]

Basic principles of mesh adaptation for parabolic FBPs

In dealing with fixed domain approximations to parabolic FBPs, the interface does not play any explicit role but, nonetheless, is responsible for the global numerical pollution on finite element solutions [8,9]. A remedy to this undesirable situation is to be found in terms of mesh adaptation [3,4,5,6]. Its basic principles are fully discussed in light of the two-phase Stefan problem in 2-D, considered here as a model example. The local refinement strategy is based on equidistributing pointwise interpolation errors as well as on specifying the so-called refined region, which in turn must contain the discrete free boundaries. This is accomplished by performing various tests on the computed temperature to extract information about its first and second derivatives as well as to predict free boundary locations. A typical triangulation is coarse away from the discrete interface, where the discretization parameters satisfy a parabolic relation, whereas it is locally refined in the vicinity of the discrete interface for the relation to become hyperbolic (refined region); a drastic reduction of spatial degrees of freedom is obtained with highly graded meshes. The local truncation error in time is thus properly balanced with the local interpolation error in space. Consecutive meshes are not compatible in that they are completely regenerated rather than being produced by enrichment-coarsening strategies. Mesh changes incorporate an interpolation error which eventually accumulates in time. A suitable interpolation theory for noncompatible meshes is developed to quantify such an error. Its control imposes several constraints on admissible meshes and leads to the mesh selection algorithm. The resulting scheme is stable in various Sobolev norms. A rate of convergence of essentially $O(\tau^{1/2})$ is derived in the natural energy spaces for both temperature and enthalpy, where $\tau > 0$ stands for the (uniform) time step.

1. Local Mesh Constraints. Let $\Omega \subset \mathbf{R}^2$ be a convex polygon and $T > 0$ be fixed. Set $\beta(s) := (s-1)^+ - s^-$; β may even be nonlinear on $\mathbf{R}\backslash[0,1]$ [4]. The two-phase Stefan problem consists of finding $\{u, \theta\}$ such that $u_t - \Delta\theta = f(\theta)$, $\theta = \beta(u)$, in $\Omega \times (0,T)$, subject to $\theta = 0$ on $\partial\Omega \times (0,T)$ and $u(0) = u_0$.

Let $\{\mathcal{S}^n\}_{n=1}^N$ denote a set of graded partitions of Ω into triangles, that are shape regular [1, p.132] and weakly acute uniformly with respect to $1 \le n \le N := T/\tau$. The second condition means that for any pair of adjacent triangles the sum of the opposite angles relative to the common side does not exceed π. Given a triangle $S \in \mathcal{S}^n$, h_S

†Department of Mathematics and Institute for Physical Science and Technology, University of Maryland, College Park, MD 20742 USA.

‡Istituto di Analisi Numerica del CNR, 27100 Pavia, Italy.

♯Dipartimento di Meccanica Strutturale, Università di Pavia, 27100 Pavia, Italy.

stands for its size and verifies $\lambda\tau \leq h_S \leq \Lambda\tau^{1/2}$, where $0 < \lambda, \Lambda$ are fixed constants. Let $\mathcal{R}^n := \cup\{S \in \mathcal{S}^n : h_S = O(\tau)\}$ indicate the *refined region*. Let $\mathbf{V}^n$ be the space of continuous piecewise linear finite elements over $\mathcal{S}^n$. Let $\Pi^n : C^0(\bar{\Omega}) \to \mathbf{V}^n$ be the usual Lagrange interpolation operator.

Let $U^0 \in \mathbf{V}^0 := \mathbf{V}^1$ be a suitable approximation to u_0 and $\Theta^0 := \Pi^1\beta(U^0)$ [4]. Given a mesh $\mathcal{S}^{n-1}$ and a discrete enthalpy $U^{n-1} \in \mathbf{V}^{n-1}$ for any $1 \leq n \leq N$, the discrete scheme then reads as follows: *select $\mathcal{S}^n$ and find $U^n, \Theta^n \in \mathbf{V}^n$ such that*

$$(1.1) \qquad\qquad \Theta^n = \Pi^n\beta(U^n),$$

$$(1.2) \qquad\qquad \hat{U}^{n-1} := \Pi^n U^{n-1}, \qquad \hat{\Theta}^{n-1} := \Pi^n[\beta(U^{n-1})],$$

$$(1.3) \qquad \tau^{-1}\langle U^n - \hat{U}^{n-1}, \chi\rangle^n + \langle\nabla\Theta^n, \nabla\chi\rangle = \langle f(\hat{\Theta}^{n-1}), \chi\rangle^n, \qquad \forall\, \chi \in \mathbf{V}^n,$$

where $\langle\varphi, \chi\rangle^n := \int_\Omega \Pi^n(\varphi\chi)dx$ for all $\varphi, \chi \in C^0(\bar{\Omega})$. The ensuing algebraic system is (strongly) nonlinear and strictly monotone because (1.1) enforces the nonlinear constitutive relation $\theta = \beta(u)$ at the discrete level. Its unique solution can be easily and efficiently computed via a nonlinear SOR method [5,7,8,9].

To fully describe the adaptive FEM we must indicate how to select the new mesh $\mathcal{S}^n$; this is a crucial task. We will define three local parameters over $\mathcal{S}^{n-1}$ which, conveniently postprocessed [5,7], give rise to a meshsize function $\mathbf{h}^n$ with which $\mathcal{S}^n$ is next generated by the automatic mesh generator of [10].

The key idea hinges upon equidistributing pointwise interpolation errors for Θ^{n-1} in order that the maximum norm error produced by a mesh change be $O(\tau)$. We will now discuss the nature of such an error but before we need some preparations. We first remove the superscript n in the various quantities involved and use, instead, the compact notation: $\mathcal{S} := \mathcal{S}^{n-1}$, $\hat{\mathcal{S}} := \mathcal{S}^n$, $\mathcal{R} := \mathcal{R}^{n-1}$, $\Pi := \Pi^{n-1}$, $\hat{\Pi} := \Pi^n$, $U := U^{n-1}$, $\hat{U} := \hat{\Pi}U$, $\Theta := \Theta^{n-1}$. The discrete interface is defined by $F := \{\Theta = 0\}$. Let $\mathcal{E}$ denote the set of interelement boundaries of $\mathcal{S}$. We then set $\mathcal{S}_W := \{S \in \mathcal{S} : S \cap W \neq \emptyset\}$ and $\mathcal{E}_W := \{e \in \mathcal{E} : e \subset \partial S, S \in \mathcal{S}_W\}$, for any given set $W \subset \bar{\Omega}$. Set $d_S := |\nabla\Theta|_S|$ for all $S \in \mathcal{S}$ and $h_e := \text{lenght}(e)$, $D_e := |[\![\nabla\Theta]\!]_e|/h_e$ for all $e \in \mathcal{E}$, where $[\![\cdot]\!]_e$ indicates the jump operator on e. Note that these quantities are easy to evaluate in practice. We then introduce the following local parameters

$$(1.4) \qquad \hat{h}_e := \mu_1 \frac{\tau^{1/2}}{D_e^{1/2}}, \quad \forall\, e \in \mathcal{E}\backslash\mathcal{E}_F; \qquad \hat{h}_S := \mu_2 \frac{\tau^{1/2}}{d_S}, \quad \forall\, S \in \mathcal{S}\backslash\mathcal{S}_F;$$

$\lambda, \mu_1, \mu_2 > 0$ are arbitrary constants which result from computational considerations [5]. Set $\mathcal{S}_B := \{S \in \mathcal{S}\backslash\mathcal{S}_F : \min_{e\in\mathcal{E}\backslash\mathcal{E}_F : e\subset\partial S}(\hat{h}_S, \hat{h}_e) < \lambda\tau\}$, $\mathcal{E}_B := \{e \in \mathcal{E}\backslash\mathcal{E}_F : e \subset \partial S, S \in \mathcal{S}_B\}$, $\mathcal{B} := \cup_{S\in\mathcal{S}_B}S$, $\mathcal{F} := \cup_{S\in\mathcal{S}_F}S$, $\mathcal{S}_0 := \mathcal{S}\backslash(\mathcal{S}_F\cup\mathcal{S}_B)$, $\mathcal{E}_0 := \mathcal{E}\backslash(\mathcal{E}_F\cup\mathcal{E}_B)$, $\Omega_0 := \Omega\backslash(\mathcal{F}\cup\mathcal{B})$. All triangles in $\mathcal{S}_B$, for which either first or second derivatives are badly-behaved, as may happen near F, are kept fixed, namely

$$(1.5) \qquad\qquad S \in \hat{\mathcal{S}}, \qquad \forall\, S \in \mathcal{S}_B.$$

178

We are now ready to comment on the heuristic idea behind (1.4). Let $\zeta : \mathbf{R} \backslash \{0\} \to \mathbf{R}$ be a smooth nonlinear function; examples relevant to our study are $\zeta(s) = \beta^{-1}(s)$ and $\zeta(s) = s^2$. The definition (1.4) is so devised that the local pointwise interpolation error corresponding to $\zeta(\Theta)$ satisfies $\hat{h}_e^2 D_e + \hat{h}_S^2 d_S^2 = O(\tau)$; see §3. Since this error accumulates in time, we impose a restriction to the maximum number of mesh changes, namely $O(\tau^{-1/2})$, for the final error to be $O(\tau^{1/2})$. We notice that second derivatives D_e (first derivatives d_S) may blow up without violating $\hat{h}_e \geq \lambda \tau$ ($\hat{h}_S \geq \lambda \tau$) as far as $D_e \leq (\mu_1/\lambda)^2 \tau^{-1}$ ($d_S \leq (\mu_2/\lambda)\tau^{-1/2}$). This is consistent with the expected values of D_e and d_S as heuristics and numerical experiments indicate [3,4,5]. They also suggest that $\sum_{e \in \mathcal{E}} h_e^2 D_e \leq D$, with $D > 0$ independent of τ, that we further assume. Such a *structural* assumption is a discrete analogue of $\Delta \theta \in L^\infty(0, T; M(\Omega))$, which is known to hold for the continuous problem; $M(\Omega)$ stands for the space of finite regular Baire measures in Ω.

The remaining local parameter represents the expected local meshsize within the refined region, and so near the (nondegenerate) discrete interface [4]. We first impose

$$(1.6) \qquad S \in \hat{\mathcal{S}}, \qquad \forall\, S \in \mathcal{S}_F,$$

to prevent the error bound $\|U - \hat{U}\|_{L^\infty(\Omega)} = O(1)$ from happening. For any $S \in \mathcal{S}_F$, we first determine an approximation to the (average) interface velocity V_S [3,4,5,6]. We then define $\mathcal{C}_S$ to be the cone of axis ν, vertex at S, opening $\pi/2$ and height $\mu_3 V_S \tau^{1/2}$ as being the region most likely to contain the evolution of $F_S := S \cap F$ for at least $O(\tau^{-1/2})$ time steps. Heuristic arguments, based on balancing the truncation error in time with the interpolation error in space [3,4,5], suggest the following correction

$$(1.7) \qquad \hat{h}_{F_S} := \tau \min\{\max(\lambda, V_S), M\},$$

for the meshsize within $\mathcal{C}_S$. The constants μ_3 and M are arbitrary and result from computational considerations [5]; μ_3 may depend on n. If mushy regions occur we then enforce (1.6) and (1.7) on the nondegenerate part of the mushy boundary only [4,5].

2. Mesh Selection Algorithm.

Various tests are performed on the computed solution Θ to verify admissibility of the current mesh $\mathcal{S}$. The first test consists of checking whether the discrete interface F is within the refined region $\mathcal{R}$ or not. In the event F has escaped from $\mathcal{R}$, both $\mathcal{S}$ and $\{\Theta, U\}$ are discarded and the previous computed solution recovered. To prevent the program from performing a useless time step, the boundary of $\mathcal{R}$, called RED ZONE, alerts that an imminent remeshing must be done. Since the local width of $\mathcal{R}$ is proportional to both $\tau^{1/2}$ and the local interface velocity, we thus expect $\mathcal{S}$ to be admissible for at least $O(\tau^{-1/2})$ time steps, as desired.

The second test ascertains that interpolation errors are still equidistributed correctly,

$$(2.1) \qquad h_e \leq \mu_1^* \hat{h}_e, \quad \forall\, e \in \mathcal{E}_0, \qquad h_S \leq \mu_2^* \hat{h}_S, \quad \forall\, S \in \mathcal{S}_0,$$

where $\mu_1^*, \mu_2^* > 1$ are suitable constants. This rules out the possibility of an excessive refinement induced by large discrete derivatives.

To guarantee a correct balance between local truncation error and meshsize near F as well as to avoid computational difficulties in specifying the refined region, the third test

$$(2.2) \qquad \mu_3^- \hat{h}_{F_S} \leq h_S \leq \mu_3^+ \hat{h}_{F_S}, \qquad \forall\, S \in \mathcal{S}_F,$$

is enforced; $\mu_3^- < 1 < \mu_3^+$ are suitable constants. If a single constraint in either (2.1) or (2.2) is violated, the current mesh $\mathcal{S}$ is rejected and the previous computed solution recovered. The remeshing is, however, mostly dictated by the discrete free boundary location [5].

3. Interpolation Theory. Our present concern is interpolation estimates for non-compatible meshes. We stick with the notation of §1. Given $W \subset \bar{\Omega}$, we set $\widetilde{W} := \cup_{S \in \mathcal{S}_W} S$ and point out that there exists a constant $0 < a < 1$, depending only on the regularity of $\mathcal{S}$, such that $\text{dist}(x, S) \geq a h_S$ for all $x \in \bar{\Omega}_0 \backslash \tilde{S}$. We can then distinguish between two opposite (and mutually exclusive) situations in terms of the relative size of triangles in both meshes $\mathcal{S}$ and $\hat{\mathcal{S}}$. Set $\hat{\mathcal{S}}_0 := \hat{\mathcal{S}} \backslash (\mathcal{S}_F \cup \mathcal{S}_B)$ and define the *derefinement* case to be

$$(3.1) \qquad \text{given } \hat{S} \in \hat{\mathcal{S}}_0, \quad a h_S < h_{\hat{S}} \quad \text{for all } S \in \mathcal{S}_{\hat{S}}.$$

By contrast, the *refinement* situation reads as follows

$$(3.2) \qquad \text{given } \hat{S} \in \hat{\mathcal{S}}_0, \quad \text{there exists } S \in \mathcal{S}_{\hat{S}} \quad \text{such that } a h_S \geq h_{\hat{S}},$$

which in turn yields $\hat{S} \subset \tilde{S}$. Let $\hat{\mathcal{S}}_1$ (resp. $\hat{\mathcal{S}}_2$) indicate the set of all $\hat{S}$'s satisfying (3.1) (resp. (3.2)). We then have the following two crucial L^p-interpolation estimates without assumptions on the relative size or location of new and old triangles [4].

LEMMA 3.1. *Let* $\hat{S} \in \hat{\mathcal{S}}_1$. *Then*

$$(3.3) \quad \|U - \hat{U}\|_{L^p(\hat{S})} + h_{\hat{S}} \|\nabla(U - \hat{U})\|_{L^p(\hat{S})} \leq C h_{\hat{S}}^2 \begin{cases} (\sum_{e \in \mathcal{E}_{\hat{S}}} h_e^2 D_e^p)^{1/p}, & 1 \leq p < \infty \\[2mm] \max_{e \in \mathcal{E}_{\hat{S}}} D_e, & p = \infty. \end{cases}$$

The need of control on interpolation errors even for the refinement situation (3.2) arises from the noncompatibility of $\mathcal{S}$ and $\hat{\mathcal{S}}$.

LEMMA 3.2. *Let* $\hat{S} \in \hat{\mathcal{S}}_2$. *Then*

$$(3.4) \quad \|U - \hat{U}\|_{L^p(\hat{S})} + h_{\hat{S}} \|\nabla(U - \hat{U})\|_{L^p(\hat{S})} \leq \begin{cases} C h_{\hat{S}}^{1+2/p} (\sum_{e \in \mathcal{E}_{\hat{S}}} h_e^p D_e^p)^{1/p}, & 1 \leq p < \infty \\[2mm] C h_{\hat{S}} \max_{e \in \mathcal{E}_{\hat{S}}} (h_e D_e), & p = \infty. \end{cases}$$

It is worth noting that both (3.3) and (3.4) may be rewritten in terms of Θ, rather than U, because they simply coincide or differ by 1 for all $\hat{S} \in \hat{\mathcal{S}}_0 (= \hat{\mathcal{S}} \backslash (\mathcal{S}_F \cup \mathcal{S}_B))$. This assertion holds even for nonlinear β [4]. The following estimates are a consequence of the mesh selection algorithm and the two preceding results [3,4,6].

180

THEOREM 3.1. *The following sharp interpolation error estimates are valid*

(3.5) $$\|U - \hat{U}\|_{L^\infty(\Omega)} \le C\tau, \qquad \|\nabla(U - \hat{U})\|_{L^2(\Omega)} \le C\tau^{1/2}.$$

Since meshes $\mathcal{S}$ and $\hat{\mathcal{S}}$ are noncompatible, we cannot expect a pointwise error estimate for ∇U to hold. In fact, consider the refinement case (i.e. $\hat{S} \in \hat{\mathcal{S}}_2$) for which $\|\nabla(U - \hat{U})\|_{L^\infty(\hat{S})} \le C \max_{e \in \mathcal{E}_{\hat{S}}} h_e D_e$. This yields $\|\nabla(U - \hat{U})\|_{L^\infty(\hat{S})} = O(1)$ provided $D_e = O(h_e^{-1})$, as expected to happen near the interface F.

4. Stability. Since $\mathcal{S}^n$ is weakly acute, the *discrete maximum principle* is valid. This serves to exploit monotonicity properties of the problem at hand which in turn compensate for the lack of regularity. As a consequence of results in §3, we have the following natural *a priori estimates* for the discrete problem [4]:

(4.1) $$\max_{1 \le n \le N} \|U^n\|_{L^\infty(\Omega)} + \max_{1 \le n \le N} \|\Theta^n\|_{L^\infty(\Omega)} \le C.$$

(4.2) $$\sum_{n=1}^{N} \|U^n - \hat{U}^{n-1}\|_{L^2(\Omega)}^2 + \sum_{n=1}^{N} \tau \|\nabla\Theta^n\|_{L^2(\Omega)}^2 \le C.$$

(4.3) $$\sum_{n=1}^{N} \tau^{-1} \|U^n - \hat{U}^{n-1}\|_{L^2(\Omega \setminus \mathcal{R}^n)}^2 + \max_{1 \le n \le N} \|\nabla\Theta^n\|_{L^2(\Omega)} \le C,$$

(4.4) $$\max_{1 \le n \le N} \|U^n - \hat{U}^{n-1}\|_{L^1(\Omega)} \le C\tau.$$

The leftmost term in (4.2) is a time-discrete $H^{1/2}$ estimate and accounts for the global behavior of U^n which, in the limit, is discontinuous. The corresponding term in (4.3) states, instead, a time-discrete H^1 regularity away from the interface F^n, where U^n and Θ^n are equivalent variables. The a priori estimate (4.4) is a discrete analogue of $u_t \in L^\infty(0, T; M(\Omega))$, but still a bit weaker than the structural assumption $\sum_{e \in \mathcal{E}} h_e^2 D_e \le D$. We stress the need for stability estimates in nonenergy spaces such as (4.1) and (4.4) as they play a relevant role in the error analysis.

5. Error Estimates. The error analysis is based upon (4.1) to (4.4) and a quasi-optimal pointwise error estimate for the Laplace operator on highly graded meshes satisfying $\max_{S \in \mathcal{S}^n} h_S \le C \min_{S \in \mathcal{S}^n} h_S^\gamma$ for $0 < \gamma \le 1$ [4]; $\gamma = 1/2$ is this context.

THEOREM 5.1. *Let the number of mesh changes be $O(\tau^{-1/2})$ at most. Then*

(5.1) $$\|e_u\|_{L^\infty(0,T;H^{-1}(\Omega))} + \|e_\theta\|_{L^2(0,T;L^2(\Omega))} \le C\tau^{1/2} |\log \tau|^{7/2}.$$

For the practical range of time steps τ, the logarithm above plays no significant role. As explained in §1, the restriction on the number of mesh changes accounts for the accumulation of interpolation error $U^{n-1} - \hat{U}^{n-1}$ which is $O(\tau)$. The major novelty in Theorem 5.1 is to be interpreted in terms of properly distributed spatial degrees of freedom (DOF): for well-behaved discrete interfaces, only DOF$=O(\tau^{-3/2})$ are necessary for an $O(\tau^{1/2})$ global accuracy, as opposed to quasi-uniform meshes that require DOF$=O(\tau^{-2})$ [2,8].

6. Quasi-Optimal Distribution of DOF. As of now, most of the DOF are concentrated near discrete interfaces for the so-called *refined region* $\mathcal{R}^n$ to be a strip $O(\tau^{1/2})$-wide. This in turn comes from restricting the number of mesh changes to $O(\tau^{-1/2})$ as a consequence of accuracy considerations. As numerical evidence indicates, most of the CPU time is spent in solving the ensuing strongly nonlinear algebraic systems rather than in regenerating or interpolating [5,7]. It would thus be preferable to have a narrower refined region, say $O(\tau)$-wide, at the expense of changing the mesh more frequently, say every $O(1)$ time steps. Hence, the resulting meshes would have DOF$=O(\tau^{-1})$, which appears to be quasi-optimal as corresponds to the expected value for the (linear) heat equation.

Let $\mathbf{W}^n$ be the space of (discontinuous) piecewise linear functions over $\mathcal{S}^n$. Let $P^n : L^2(\Omega) \to \mathbf{W}^n$ be the (local) L^2-projection operator defined by $\langle P^n v, \chi \rangle := \langle v, \chi \rangle$ for all $v \in L^2(\Omega), \chi \in \mathbf{W}^n$. Since no continuity requirements are imposed on $\mathbf{W}^n$, $P^n v$ can be computed elementwise by inverting a 3×3 linear system. Let $\{x_j^n\}_{j=1}^{J^n}$ denote the nodes of $\mathcal{S}^n$ and $\{\chi_j^n\}_{j=1}^{J^n}$ the canonical basis of $\mathbf{V}^n$; $\mathrm{supp}\chi_j^n = \cup_{S \in \mathcal{S}_{x_j^n}^n} S$. Let now $\hat{U}^{n-1} \in \mathbf{V}^n$ be

$$(6.1) \qquad \hat{U}^{n-1}(x_j^n) := \sum_{S \in \mathcal{S}_{x_j^n}^n} \frac{\mathrm{meas}\,(S)}{\mathrm{meas}\,(\mathrm{supp}\chi_j^n)} [P^n U^{n-1}]|_S(x_j^n), \qquad \forall\, 1 \le j \le J^n,$$

and $\hat{\Theta}^{n-1} := \Pi^n(\beta(\hat{U}^{n-1}))$. Then $\|U^{n-1} - P^n U^{n-1}\|_{H^{-1}(\Omega)} \le C\tau^{3/2}$, which is a superconvergence estimate in the space in which to measure accuracy for enthalpy. The accumulation of interpolation errors in $H^{-1}(\Omega)$ is thus $O(\tau^{1/2})$ even for $O(\tau^{-1})$ mesh changes. All the previous results remain valid if we allow an $\mathcal{R}^n$-width of $O(\tau)$, as desired, and enforce the quality tests of §2 [6]. However, the implementation of P^n is far from trivial and intrinsically more expensive than that of Π^n. This is subject of current research.

REFERENCES

[1] P.G. CIARLET, *The finite element method for elliptic problems*, North Holland, Amsterdam, 1978.

[2] C.M. ELLIOTT, *Error analysis of the enthalpy method for the Stefan problem*, IMA J. Numer. Anal., 7 (1987), pp. 61–71.

[3] R.H. NOCHETTO, M. PAOLINI AND C. VERDI, *Selfdaptive mesh modification for parabolic FBPs: theory and computation*, in Proceedings Free boundary value problems with special respect to their numerical treatment and optimal control, Birkhäuser, Stuttgart, 1989, to appear.

[4] R.H. NOCHETTO, M. PAOLINI AND C. VERDI, *An adaptive finite element method for two-phase Stefan problems in two space dimensions. Part I: Stability and error estimates*, Math. Comp. (to appear).

[5] R.H. NOCHETTO, M. PAOLINI AND C. VERDI, *An adaptive finite element method for two-phase Stefan problems in two space dimensions. Part II: Implementation and numerical experiments*, (to appear).

[6] R.H. NOCHETTO, M. PAOLINI AND C. VERDI, *Quasi-optimal mesh adaptation for two-phase Stefan problems in 2D*, in Computational Mathematics and Applications, Istituto di Analisi Numerica del C.N.R., 730, Pavia, Italy, 1989, pp. 313–326.

[7] R.H. NOCHETTO, M. PAOLINI AND C. VERDI, *Efficient mesh adaptation for parabolic FBPs*, this conference (to appear).

[8] R.H. NOCHETTO AND C. VERDI, *Approximation of degenerate parabolic problems using numerical integration*, SIAM J. Numer. Anal., 25 (1988), pp. 784–814.

[9] M. PAOLINI, G. SACCHI AND C. VERDI, *Finite element approximations of singular parabolic problems*, Int. J. Numer. Meth. Eng., 26 (1988), pp. 1989–2007.

[10] M. PAOLINI AND C. VERDI, *An automatic mesh generator for planar domains*, Rivista di Informatica (to appear).

R H NOCHETTO[†], M PAOLINI[‡] AND C VERDI[#]
Efficient mesh adaptation for parabolic FBPs

Numerical pollution associated with the lack of regularity of solutions across the interfaces makes local refinement methods very attractive for parabolic FBPs. This contribution addresses various implementation aspects that are essential for such methods to be competitive. The two-phase Stefan problem in 2D is viewed as a model example. Several numerical tests are performed on the computed temperature to extract information about its discrete regularity as well as to predict free boundary locations. Whenever the current mesh is to be discarded, either because the discrete interface escapes from the so-called refined region or interpolation errors are no longer properly distributed, a new mesh is generated. Following the modern trend for transient problems, remeshing of the whole domain is used in place of enrichment-coarsening strategies, thus making mesh generator efficiency an essential issue. The discrete enthalpy, which is piecewise linear in the old mesh, is then interpolated in the new mesh to advance the algorithm in time. This is another crucial process because consecutive meshes are *noncompatible* and highly *graded*. Binary search techniques on suitable quadtree structured data are used to reach a quasi-optimal computational complexity in several search operations in both generation and interpolation algorithms. The proposed method necessitates much less spatial degrees of freedom than previous practical methods on quasi-uniform meshes to achieve the same global asymptotic accuracy. This is reflected in its superior performance as expressed in terms of computing time for a desired accuracy. It is also robust in that it handles degenerate situations such as the occurrence of phases and mushy regions as well as the formation of cusps. Several numerical experiments illustrate those properties along with the scheme efficiency in approximating both solutions and interfaces in maximum norm.

1. The Adaptive FEM. Let $\Omega \subset \mathbf{R}^2$ be a convex polygon and $T > 0$ be fixed. Set $\beta(s) := (s-1)^+ - s^-$. The two-phase Stefan problem consists of finding $\{u, \theta\}$ such that $u_t - \Delta\theta = f(\theta)$, $\theta = \beta(u)$, in $Q := \Omega \times (0, T)$, subject to $u(0) = u_0$ and boundary conditions for θ on $\partial\Omega \times (0, T)$.

Let $\tau := T/N$ be the time step and $\mathcal{S}^n$ a partition of Ω into triangles; $\mathcal{S}^n$ is assumed to be *weakly* acute and *regular* uniformly in $1 \le n \le N$, [1,p.132;5]. For any $S \in \mathcal{S}^n$, h_S stands for its size and verifies $\lambda\tau \le h_S \le \Lambda\tau^{1/2}$ ($0 < \lambda, \Lambda$ fixed). Let $\mathcal{R}^n := \cup\{S \in \mathcal{S}^n : h_S = O(\tau)\}$ denote the *refined region*. Let $\mathbf{V}^n \subset H_0^1(\Omega)$ indicate the usual piecewise linear finite element space over $\mathcal{S}^n$ and $\Pi^n : C^0(\bar{\Omega}) \to \mathbf{V}^n$ the Lagrange interpolation operator.

---

†Department of Mathematics and Institute for Physical Science and Technology, University of Maryland, College Park, MD 20742 USA.

‡Istituto di Analisi Numerica del CNR, 27100 Pavia, Italy.

#Dipartimento di Meccanica Strutturale, Università di Pavia, 27100 Pavia, Italy.

Let $U^0 \in \mathbf{V}^0 := \mathbf{V}^1$ be a suitable approximation to u_0 and $\Theta^0 := \Pi^1\beta(U^0)$ [3]. Given a mesh $\mathcal{S}^{n-1}$ and a discrete enthalpy $U^{n-1} \in \mathbf{V}^{n-1}$ for any $1 \leq n \leq N$, the discrete scheme reads as follows: *select $\mathcal{S}^n$ and find $U^n, \Theta^n \in \mathbf{V}^n$ such that $\Theta^n = \Pi^n\beta(U^n)$ and*

$$(1.1) \quad \int_\Omega \Pi^n((U^n - U^{n-1})\chi) + \tau \int_\Omega \nabla\Theta^n \cdot \nabla\chi = \tau \int_\Omega \Pi^n(f(\beta(U^{n-1}))\chi), \qquad \forall \chi \in \mathbf{V}^n.$$

Let $\mathbf{M}^n$ be the resulting diagonal mass matrix and $\mathbf{K}^n$ the stiffness matrix. Then,

$$(1.2) \qquad \mathbf{M}^n U^n + \tau\mathbf{K}^n\Theta^n = \mathbf{M}^n(\hat{U}^{n-1} + \tau\hat{Q}^{n-1})$$

is a matrix formulation equivalent to (1.1), where we have identified piecewise linear functions with the vector of their nodal values. Note that $\hat{U}^{n-1} := \Pi^n U^{n-1}$ is the interpolant of $U^{n-1} \in \mathbf{V}^{n-1}$ in the new mesh $\mathcal{S}^n$, which is not compatible with $\mathcal{S}^{n-1}$, and that $\hat{Q}^{n-1} := \Pi^n[f(\beta(U^{n-1}))]$ [3,4,5]. The strongly nonlinear algebraic system (1.2) is efficiently solved by an optimized nonlinear SOR method [4;6,7]; see §3.

The basic principles of mesh adaptation are discussed in [3,5]. A typical mesh is designed to be coarse away from the discrete interface $F^n := \{\Theta^n = 0\}$, where discretization parameters satisfy the parabolic relation $h_S = O(\tau^{1/2})$, and locally refined in the vicinity of F^n for the relation to become hyperbolic $h_S = O(\tau)$; see Figs 3.1 and 3.2.

2. Computational Issues. Various tests are performed on the computed temperature Θ^{n-1} by the routine TEST to verify admissibility of the current mesh $\mathcal{S}^{n-1}$. First, the algorithm checks whether the discrete interface F^{n-1} lies in the interior of the refined region $\mathcal{R}^{n-1}$ (TEST=OK) or not. In the event F^{n-1} escapes from $\mathcal{R}^{n-1}$ (TEST=FAIL), the computed solution U^{n-1} is discarded and the previous one recovered; $\mathcal{S}^{n-1}$ is then rejected. To prevent the program from performing a useless time step, triangles along $\partial\mathcal{R}^{n-1}$ (blackened elements in Fig 3.2) alert that an imminent remeshing must be done (TEST=ALERT). The second test ascertains that pointwise interpolation errors are still equidistributed correctly (TEST=OK) in order to rule out the possibility of an excessive refinement induced by large discrete derivatives. A third test finally verifies whether the local truncation error and meshsize match near the discrete interface (TEST=OK) or not.

If TEST=OK, then $\mathcal{S}^n := \mathcal{S}^{n-1}$ is selected; otherwise a new mesh has to be generated. The expected local meshsize, extracted from the tests above, is stored in a piecewise constant function $\check{h}^n$ defined on a fixed auxiliary uniform square mesh $\mathcal{Q}$ of size $O(\tau^{1/2})$. This function is postprocessed by the routine H_DEFINE, which constructs another piecewise constant function $\mathbf{h}^n$ satisfying $\mathbf{h}^n|_R \leq \check{h}^n|_P$ for all $P \in \mathcal{Q}$ such that $\mathrm{dist}(P, R) \leq \mathbf{h}^n|_R$. The automatic mesh generator MESH of [8] next uses $\mathbf{h}^n$ to produce an adequate (weakly acute) triangulation $\mathcal{S}^n$, namely one verifying the compatibility constraint: $h_S \leq \check{h}^n(x)$ for all $x \in S$, where S is any element in $\mathcal{S}^n$.

The FE code was written in C-language, on a VAX 8530 VMS 5.1, because of the need to avoid very large static memory space allocations and also use *quadtrees*. Such data

184

structures are in essence a convenient organization of a given finite set of points of $\mathbf{R}^2$, that in turn allows several search operations to be performed with logarithmic complexity. They are essentially to find, insert and remove a point, to find a nearest point to a given one and to find all points that belong to a given set. Quadtrees are then used in both the advancing front algorithm of MESH, that starting from the fixed boundary invades Ω, and the nontrivial interpolation process $\hat{U}^{n-1} = \Pi^n U^{n-1}$ carried out by INTERPOLATION. Implementation details can be found in [4,8]. A schematic flow chart of the program is as follows:

$n = 1$; do (while $n \leq N$)

{H_DEFINE; MESH   (defines $\mathbf{h}^n$ and generates $\mathcal{S}^n$)

 MATRIX   (computes $\mathbf{M}^n$ and $\mathbf{K}^n$)

 INTERPOLATION   (computes $\hat{U}^{n-1} = \Pi^n U^{n-1}$)

 do {SOR; $n = n + 1$} while (TEST=OK & $n \leq N$)   (computes U^n and Θ^n)

 if (TEST=FAIL) {$n = n - 1$; SOL_UPDATE}}   (retrieves U^{n-1})

The main advantage of our refinement strategy is its computational simplicity and efficiency, as CPU times for the various routines reveal; see §3. In particular a mesh data structure is eliminated as soon as the interpolation process has been carried out, thus releasing the corresponding memory allocation. On the contrary, enrichment/coarsening procedures require a delicate data structure to keep track of successive mesh changes, which in turn seems to be very sensitive to the number of refinement levels. For strongly graded meshes, such as those here, this becomes a critical issue. The computational complexity of the various tasks is ($J^n :=$ number of nodes of $\mathcal{S}^n$): H_DEFINE, MATRIX, TEST and SOL_UPDATE $\rightarrow$CC$_n = O(J^n)$; MESH and INTERPOLATION $\rightarrow$ CC$_n = O(J^n \log J^n)$; SOR $\rightarrow$ CC$_n = O((J^n)^{4/3} \log N)$. Note that CC$_n$ for both MESH and INTERPOLATION is *quasi-optimal*.

The refined region $\mathcal{R}^n$ is typically a strip $O(\tau^{1/2})$-wide around the polygonal F^n; in the presence of mushy regions or wildly behaved interfaces, $\mathcal{R}^n$ might be much bigger though. Since its local meshsize is $O(\tau)$, the number of triangles within $\mathcal{R}^n$ is $O(N^{3/2})$ as opposed to $O(N)$ for the number of triangles outside of $\mathcal{R}^n$. The resulting number of spatial degrees of freedom is thus $J^n = O(N^{3/2})$. Hence, the total computational complexity (TCC) turns out to be TCC= $O(N^3 \log N)$ for a global accuracy $O(N^{-1/2})$. Since the number of mesh changes is typically $O(N^{1/2})$, the total overhead of our code (work not related to solving (1.2)) is $O(N^2 \log N)$, which is asymptotically negligible compared to TCC. Using a fixed quasi-uniform mesh, instead, would lead to TCC= $O(N^{11/3} \log N)$ for the same global accuracy because $h_S = O(\tau)$ for all $S \in \mathcal{S}^n$ [2,6]. An improvement, though not so dramatic, is also gained for mushy regions as shown in Table 3.3.

3. Numerical Experiments. Several numerical experiments were performed to illustrate the superior performance of our Adaptive FEM (AM) [4] with respect to the standard one with a fixed mesh (FMM) [2,6,7]. The first experiment corresponds to the evolution

185

of a smooth interface, the second example illustrates the formation of a cusp whereas the last one shows the evolution of a mushy region that eventually becomes a sharp interface. In presenting the numerical results, we have employed the following notation: J :=average number of nodes, $E_\theta^\infty := \|\theta - \Theta\|_{L^\infty(Q)}$, scaled $\times 10^2$, E_I^∞ :=distance between continuous and discrete interfaces, scaled $\times 10^2$, t := total CPU time in seconds and t_H, t_{MESH}, t_{MAT}, t_{TEST}, t_{INT}, t_{SOR} :=CPU time of H_DEFINE, MESH, MATRIX, TEST, INTERPOLATION, SOR, respectively, in seconds. For details we refer to [4].

3.1. *Example I: Oscillating Interface.* This is a classical two-phase Stefan problem with an interface that moves up and down. Here $\Omega := (0,5) \times (-0.5, 4.5)$, $T := \pi/1.25$ and $x^2 + (y - \sin(1.25t))^2 = 1$ is the exact interface. This example is an extremely difficult test for our numerical method, because the velocity normal to the interface exhibits a significant variation along the front. Table 3.1 shows the CPU time distribution for AM; $\|\theta\|_{L^\infty(Q)} \approx 13.38$. Table 3.2 compares t with the final pointwise accuracy for both AM and FMM. A drastic improvement is observed here as well as in Figs 3.2 and 3.3, which exhibit the location of discrete and exact interfaces for both AM and FMM, respectively.

N	J	E_θ^∞	E_I^∞	t_H	t_{MESH}	t_{MAT}	t_{TEST}	t_{INT}	t_{SOR}
40	339	10.7	5.83	7.1	16.3	1.8	6.9	1.6	15.6
60	592	7.42	5.01	14.7	38.6	3.8	17.1	4.6	46.3
80	818	5.17	3.26	25.1	68.4	6.6	24.7	8.9	99.3
120	1406	3.92	2.66	51.7	149.7	13.7	58.2	21.0	284.9
160	2110	3.44	2.32	70.8	221.5	20.6	105.3	34.2	609.3
240	3631	2.58	1.72	150.8	534.4	48.9	221.3	89.2	1855.9

TABLE 3.1. Example I: Adaptive Method.

Adaptive Method				Fixed Mesh Method			
$N \times J$	E_θ^∞	E_I^∞	t	$N \times J$	E_θ^∞	E_I^∞	t
40×339	10.7	5.83	50	100×1812	12.4	6.88	292
60×592	7.42	5.01	126	150×4107	7.78	5.65	919
80×818	5.17	3.51	235	200×7361	6.31	3.53	2264

TABLE 3.2. Example I: Comparison of Pointwise Accuracy.

3.2. *Example II: Formation of a Cusp.* This a two-phase problem with unknown exact solution and $\Omega := (-2, 4) \times (0, 5)$, $T := 1$. A cusp is expected to develop at $(0,0)$ as numerical simulation corroborates. Fig 3.4 shows the discrete interface for a number of time steps for both AM and FMM, whereas the cusp formation is depicted in Fig 3.5.

3.3. *Example III: Evolution of a Mushy Region.* This is a two-phase Stefan problem with an initial mushy region expanding in the solid phase. The enthalpy in the mush

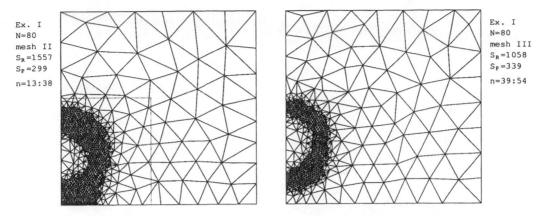

FIGURE 3.1. Example I (AM, $N = 80$): Two Consecutive Meshes.

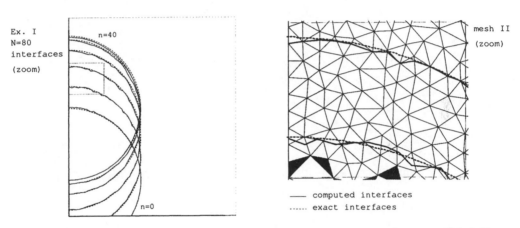

FIGURE. 3.2. Example I (AM, $N = 80$): Interfaces at $n = 8k$ ($0 \leq k \leq 5$); zoom of Mesh II.

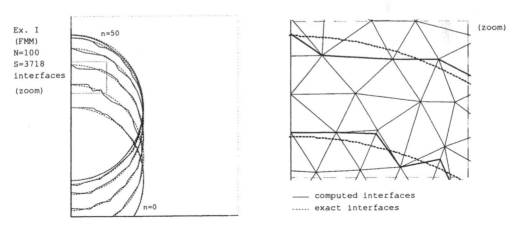

FIGURE 3.3. Example I (FMM, $N = 100$): Interfaces at $n = 10k$ ($0 \leq k \leq 5$); zoom of the Mesh.

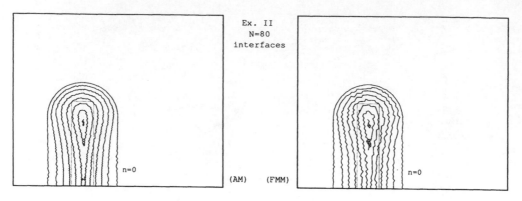

FIGURE 3.4. Example II (AM (FMM), $N = 80$): Interfaces at $n = 0, 10, 20, 30, 40, 50, 60, 67(68)$.

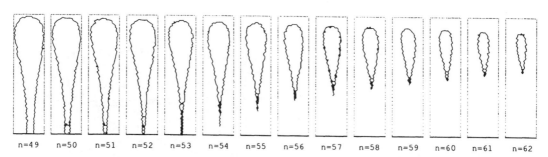

n=49 n=50 n=51 n=52 n=53 n=54 n=55 n=56 n=57 n=58 n=59 n=60 n=61 n=62

FIGURE 3.5. Example II (AM, $N = 80$): Cusp Formation (zoom).

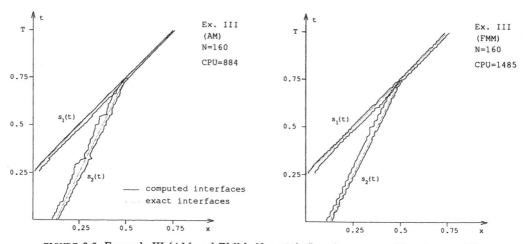

FIGURE 3.6. Example III (AM and FMM, $N = 160$): Interfaces $x = s_1(t)$ and $x = s_2(t)$.

increases so as to cause the spontaneous appearance of a liquid phase in its interior. Such a phase then expands faster than the mushy zone, which eventually disappears for the interface to become sharp and the problem a classical one. The mushy region is determined by $s_1(t) := \max(t - 0.25, 0)$ and $s_2(t) := \max(0.5(t + 0.25), t - 0.25)$ with $0 < t < T := 1$. The problem is solved in $\Omega := (0,5) \times (0,1)$ without taking any advantage of its 1D structure. We can thus compare the true interfaces $x = s_1(t)$ and $x = s_2(t)$ with the computed ones as illustrated in Fig 3.6. Pointwise accuracy is now comparable for both methods, but AM is certainly faster (see Table 3.3) and provides a better approximation of the nondegenerate interface $s_1(t)$; here $\|\theta\|_{L^\infty(Q)} \approx 6.13$.

Adaptive Method				Fixed Mesh Method			
$N \times J$	E_θ^∞	E_I^∞	t	$N \times J$	E_θ^∞	E_I^∞	t
80×506	9.09	5.47	173	80×1396	9.27	9.53	180
160×1385	1.15	4.36	884	160×5216	1.67	4.39	1485

TABLE 3.3. Example III: Comparison of Pointwise Accuracy.

4. Conclusions. The efficiency of AM is dictated by the nonlinear SOR solver, by several search operations to be performed essentially in remeshing the domain and interpolating the discrete enthalpy, and by the various quality tests. The use of quadtree structured data is thus crucial to achieve a quasi-optimal computational complexity in those search tasks. AM requires less computational effort for a desired pointwise accuracy than FMM. The improvement gained in L^∞ is extremely pronounced, as expected for a pointwise interpolation strategy. Nondegenerate exact free boundaries are approximated within one single element, thus providing a practical $O(\tau)$-rate of convergence in distance for interfaces, the best one can hope for. It seems that AM provides a powerful numerical tool to be exploited, for instance, in investigating other phase transition processes.

REFERENCES

[1] P.G. CIARLET, *The finite element method for elliptic problems*, North Holland, Amsterdam, 1978.
[2] C.M. ELLIOTT, *Error analysis of the enthalpy method for the Stefan problem*, IMA J. Numer. Anal., 7 (1987), pp. 61–71.
[3] R.H. NOCHETTO, M. PAOLINI AND C. VERDI, *An adaptive finite element method for two-phase Stefan problems in two space dimensions. Part I: Stability and error estimates*, Math. Comp. (to appear).
[4] R.H. NOCHETTO, M. PAOLINI AND C. VERDI, *An adaptive finite element method for two-phase Stefan problems in two space dimensions. Part II: Implementation and numerical experiments*, (to appear).
[5] R.H. NOCHETTO, M. PAOLINI AND C. VERDI, *Basic principles of mesh adaptation for parabolic FBPs*, this conference (to appear).
[6] R.H. NOCHETTO AND C. VERDI, *Approximation of degenerate parabolic problems using numerical integration*, SIAM J. Numer. Anal., 25 (1988), pp. 784–814.
[7] M. PAOLINI, G. SACCHI AND C. VERDI, *Finite element approximations of singular parabolic problems*, Int. J. Numer. Meth. Eng., 26 (1988), pp. 1989–2007.
[8] M. PAOLINI AND C. VERDI, *An automatic mesh generator for planar domains*, Rivista di Informatica (to appear).

D A TARZIA

Approximate and analytic methods to solve some parabolic free boundary problems

By using approximate and analytic methods we obtain an answer to some parabolic free boundary problems [Ta2] :

I. Generalized Lame'−Clapeyron solution for a one−phase source Stefan problem

Goal : "We give a generalized Lamé−Clapeyron solution for a one−phase Stefan problem with a particular type of sources. Necessary and sufficient conditions are given in order to characterize the source term which provides a unique solution".

We consider the following singular free boundary problem for the heat equation : Find the free boundary $x = s(t) > 0$, defined for $t > 0$ and $s(0) = 0$, and the temperature $\theta = \theta(x,t) > 0$, defined $0 < x < s(t)$, $t > 0$, such that they satisfy the following conditions [MeTa]:

$$\rho \, c \, \theta_t - k \, \theta_{xx} = \frac{\rho \, h}{t} \, \beta(x/2a\sqrt{t}) \quad , \quad 0 < x < s(t) \ , \ t > 0 \ ,$$

(1) $\qquad \theta(0,t) = B > 0 \ , \quad t > 0 \ , \qquad s(0) = 0 \ ,$

$\qquad \theta(s(t),t) = 0 \ , \quad k \, \theta_x(s(t),t) = - \, \rho h \, \dot{s}(t) \ , \quad t > 0 \ .$

THEOREM 1: An explicit solution of (1), as function of β, is given by

$$\theta(x,t) = B \left\{ 1 - \frac{\sqrt{\pi}}{\text{Ste}} \, \xi \, \exp(\xi^2) \, \text{erf}(\eta) + \right.$$

(2)

$$\left. + \frac{4}{\text{Ste}} \int\limits_{0}^{\eta} \left(\int\limits_{r}^{\xi} \beta(y) \, \exp(y^2) \, dy \right) \exp(-r^2) \, dr \right\} \ ,$$

$$s(t) = 2 \, a \, \xi \, \sqrt{t} \ , \qquad\qquad \eta = \ x/(2 \, a \, \sqrt{t}) \in (0 \, , \, \xi) \, , \ a^2 = \frac{k}{\rho \, c} > 0 \ ,$$

where the number $\xi > 0$ is a solution of the equation :

(3) $\qquad F(x,\beta) = \dfrac{\text{Ste}}{\sqrt{\pi}} \ , \quad x > 0 \ , \quad \text{Ste} = \dfrac{B \, c}{h} > 0 \ (\text{ Stefan number}),$

with an appropriate function $F = F(x,\beta)$.

190

THEOREM 2: Let β be a continuous real function on $\mathbb{R}^+$ such that $x\beta(x)$ is locally integrable on $\mathbb{R}^+$. Define the function Z by :

$$Z = Z_\beta(x) = \exp(x^2) \, \mathrm{erf}(x) \, [\, \psi_0(x) - \beta(x) \,] \,, \, x > 0 \,,$$

(4)

$$\psi_0(x) = \tfrac{1}{2} + x^2 + \tfrac{x}{\sqrt{\pi}} \, G(x) \,, \qquad\qquad G(x) = \frac{\exp(-x^2)}{\mathrm{erf}(x)} \,.$$

If the function Z satisfies the following conditions

(5) $\qquad Z(x) > 0 \,, \, \forall \, x \in (\nu,+\infty) \qquad$ and $\qquad \displaystyle\int_0^{+\infty} Z(t) \, dt = +\infty \,,$

where $\nu = \nu_Z \geq 0$ is defined by

(6) $\qquad \nu = \mathrm{Inf} \left\{ x \geq 0 \,/\, \displaystyle\int_0^x Z(t) \, dt > 0 \right\} \,,$

then for any Ste > 0, there exists a unique $\xi = \xi(\mathrm{Ste}) > 0$ which is the solution of the equation (3) for the given function β. Conversely, if for the given function β the equation (3) has a unique root $\xi = \xi(\mathrm{Ste}) > 0$ for any Ste>0 then there exists a continuous and locally integrable function Z on $\mathbb{R}^+$ satisfying (5) and (6) such that

(7) $\qquad \beta(x) = \psi_0(x) - Z(x) \, G(x) \quad (= \beta_Z(x)) \,, \, x > 0 \,.$

Moreover in any case the root $\xi > \nu$.

II. Neumann—like solution for the two—phase Stefan problem with a simple mushy zone model

Goal : "We give a generalized Neumann solution for a simple mushy zone model with two parameters for the two—phase Stefan problem for a semi—infinite material with equal mass densities in both solid and liquid phases, and constant thermal coefficients".

We consider a semi—infinite material with mass density equal in both solid and liquid phases and the phase—change temperature at 0°C. We generalize the mushy zone model given for the one—phase Lamé—Clapeyron (Stefan) problem in [SoWiAl] (See also [Ta1]) to the two—phase case [Ta5]. Three distinct regions can be distinguished, as follows :
H_1)The liquid phase, at temperature $\theta_2 = \theta_2(x,t) > 0$, occupying the region $x > r(t)$, $t > 0$.
H_2)The solid phase, at temperature $\theta_1 = \theta_1(x,t) < 0$, occupying the region $0 < x < s(t)$, $t > 0$.
H_3) The mushy zone, at temperature 0 , occupying the region $s(t) < x < r(t)$, $t > 0$. We make two assumptions on its structure following the paraffin case [SoWiAl] (the parameter ϵ and γ are characteristics of the phase—change material) :
a) The material in the mushy zone contains a fixed fraction ϵh (with constant $0 < \epsilon < 1$) of the total latent heat h .
b) The width of the mushy zone is inversely proportional (with constant $\gamma > 0$) to the temperature gradient at the point $(s^-(t),t)$.

THEOREM 1: If the phase—change semi—infinite material is initially in liquid phase at the constant temperature $\theta_0 > 0$ and a constant temerature $- D < 0$ is imposed on the fixed face $x = 0$, then we obtain the following results :

(i) We obtain an exact solution of the Neumann type for $\theta_1(x,t)$, $\theta_2(x,t)$, $s(t)$ and $r(t)$ as functions of θ_0 , D, ϵ , γ and the thermal coefficients of the material.

Moreover , If we replace the constant temperature $- $ D<0 by a heat flux of the type $q_0/\sqrt{t}$ (with $q_0 > 0$) on the fixed face $x = 0$, then we obtain the results :
(ii) There exists an exact solution $\theta_1^*(x,t)$, $\theta_2^*(x,t)$, $s^*(t)$ and $r^*(t)$ of the Neumann type of the mushy zone model if and only if the coefficient q_0 satisfies the inequality

$$(1) \qquad\qquad q_0 > \frac{\gamma \, k_1}{2 \, a_2 \, \eta_0} \quad ,$$

where $\eta_0 = \eta_0(\epsilon,\gamma,\theta_0,h,k_1,k_2,c_2)$ is the unique positive zero of a given function G.

Moreover, for the solution given in (i), the inequality for q_0 turns into an inequality for σ, where $\sigma > 0$ is the coefficient that characterizes the first free boundary $s(t) = 2 \, \sigma \, \sqrt{t}$ of the two$-$phase mushy zone model.

III. A new proof of the exponentially fast asymptotic behavior of the solutions in heat conduction problems with absorption

Goal : "We give a new proof of the exponentially fast asymptotic behavior of the solutions in heat conduction problems with absorption by using a variant of the heat balance integral method".

We consider the following heat conduction problem with absorption [St] :

$$\text{i) } u_t - u_{xx} + \lambda^2 \, u_+^p = 0 \; , \; x > 0 \; , \; t > 0 \; ,$$
(1)
$$\text{ii) } u(0,t) = 1 \; , \; t > 0 \; , \qquad \text{iii) } u(x,0) = U_0(x) \geq 0 \; , \; x > 0 \; ,$$

for a class of functions $U_0 = U_0(x)$ and parameters $p > 0$ and $\lambda > 0$. If $0 < p < 1$, equation (1i) has a stationary solution corresponding to datum (1ii), which has compact support in $[0,+\infty)$ and is given by :

$$(2) \qquad\qquad u_\infty(x) = \left(1 - \frac{\lambda}{I} \, x\right)_+^{\frac{2}{1-p}} \quad , \quad I = I(p) = \frac{\sqrt{2(1+p)}}{1 - p} \quad .$$

In the case $0 < p < 1$ and $U_0 \leq u_\infty$, the solution $u = u(x,t)$ of (1) satisfies

$$(3) \qquad\qquad 0 < u(x,t) < u_\infty(x) \; , \; 0 < x < \frac{I}{\lambda} \, , \, t > 0 \; ,$$

and it means that $u(t) = u(.;t)$ has compact support in variable x for any $t > 0$ and

$$(4) \qquad\qquad s(t) = \text{Sup} \, \{ \, x > 0 \, / \, u(x,t) > 0 \, \} \; , \; t > 0 \; ,$$

is a free boundary which is moving with finite speed for $t > 0$.

We give an estimate of how fast the free boundary s(t) tends to its limit I/λ as $t \to +\infty$ [Ta3,Ta4]. The estimate we get implies that this convergence is exponentially fast in time, in a similar form to the one given in [RiTa]. Our purpose is to show how this result can be obtained in a different way to [RiTa] by using the Goodman heat balance integral method [Go]. To prove that we use the innovation property (7) which fixes appropriately the asymptotic limit of the corresponding approximate free boundary. We consider a related problem to (1) which consists in finding the function

192

$C=C(x,t)$ and the free boundary $s=s(t)$ such that they satisfy the conditions:

i) $\quad C_t - C_{xx} + \lambda^2\, C_+^p = 0\ ,\ \ 0 < x < s(t)\ ,\ t > 0\ ,$

(5) ii) $\quad C(0,t)=1\ ,\ t > 0\ ,$ $\qquad$ iii) $s(0)=0\ ,$

iv) $\quad C(s(t),t) = 0\ ,\ t > 0\ ,$ $\qquad$ v) $C_x(s(t),t) = 0\ ,\ t > 0\ .$

Taking into account the heat balance integral method we replace equation (5i) by its integral in the variable x from 0 to $s(t)$, we propose

(6) $$C_B(x,t) = (1 - \frac{x}{s_B(t)})_+^\alpha$$

where $s_B = s_B(t)$ is a function to be determined and $\alpha > 1$ is a parameter to be chosen so that

(7) $$\lim_{t \to \infty} s(t) = \frac{I(p)}{\lambda}\ ,$$

then we obtain

(8) $\quad s_B(t) = \frac{I}{\lambda}\, [\, 1 - \exp(-\,\frac{2\lambda^2(3-p)t}{1+p}\,)\,]^{1/2}\ ,\ \ t \geq 0,\quad \alpha = \alpha(p) = \frac{2}{1-p} > 2.$

THEOREM 1. Let $0 < p < 1$, $\lambda > 0$ and $0 \leq U_0 \leq u_\infty$ in $\mathbb{R}^+$ be. If $u=u(x,t)$ is a solution of (1) and $s=s(t)$ is defined by (4), we have the comparison properties:

(9) $\qquad u_1(x,t) \leq u(x,t) \leq u_\infty(x)\ ,\ \ 0 \leq x \leq \frac{I}{\lambda}\ ,\ t > 0\ ,$

(10) $\qquad s_1(t) \leq s(t) \leq \frac{I}{\lambda}\ ,\ \ 0 < \frac{I}{\lambda} - s(t) \leq \frac{I}{\lambda} - s_1(t) \leq \frac{I}{\lambda}\exp(-\frac{2\lambda t}{I})\ ,\ t \geq 0\ ,$

where $u_1(x,t)$ and $s_1(t)$ are appropriate functions.

THEOREM 2. For the case $\lambda = 1$, $0 < p < 1$ and $0 \leq U_0 \leq u_\infty$ in $\mathbb{R}^+$ in problem (1), we obtain the following estimates:

(11) $\qquad s_1(t) < s_0(t)\ \leq s(t) \leq I\ ,\quad s_1(t) < s_0(t)\ < s_B(t) < I\ ,\ t > 0\ ,$

(12) $\qquad u_1(x,t) \leq u_0(x,t) \leq u(x,t) \leq u_\infty(x),\ \ u_1(x,t) \leq u_0(x,t) \leq C_B(x,t) \leq u_\infty(x), 0 \leq x \leq I,\ t > 0,$

where functions s_0 and u_0 are defined by taking $L_0 = 0$ and $m = 1$ in [RiTa].

IV. On the free boundary problem in the Wen–Langmuir shrinking core model for noncatalytic gas–solid reactions

Goal : "We give a local result in time for the existence and uniqueness of the solution of the free boundary problem in the shrinking core model for noncatalytic gas–solid reactions. We impose free boundary conditions which generalize Wen and Langmuir conditions".

193

We analyze a mathematical model of an isothermal noncatalytic diffusion—reaction process of a gas A with a solid slab S. The solid has a very low permeability and semi—thickness R along the gas diffusion direction[TaVi]. We assume the solid is chemically attacked from the surface $y = R$ with a quick and irreversible reaction of order $\nu > 0$ with respect to the gas A and zero order with respect to the solid S. We also assume that the solid is uniform and constant composition. As a result of the chemical reaction an inert layer is formed which is permeable to the gas and the process will exhibit a free boundary (the reaction front) as described in [We]. The corresponding mathematical scheme (Wen's model) is formulated as follows (in a dimensionless form) :

$$\text{i)} \quad u_{xx} - u_t = 0 \quad \text{in } D_T = \{(x, t) \,/\, 0 < x < s(t) \,,\, 0 < t \le T \},$$

$$\text{ii)} \quad u(0, t) = v_0 > 0 \,,\, 0 < t \le T \,,$$

(1)

$$\text{iii)} \quad u_x(s(t), t) = -u^{\nu}(s(t), t) \,, \qquad \text{iv) } u_x(s(t), t) = -\dot{s}(t) \,,\, 0 < t \le T \,,$$

$$\text{v)} \quad s(0) = b > 0 \,, \qquad\qquad \text{vi) } u(x, 0) = \Psi(x) \,,\, 0 \le x \le b \,.$$

We can consider the following generalized free boundary conditions :

(2)   i) $u_x(s(t), t) = g(u(s(t), t))$, ii) $\dot{s}(t) = f(u(s(t), t))$, $\quad 0 < t \le T$,

where f and g are real functions which satisfy

(3)   i) $f > 0$, $f' > 0$ in $\mathbb{R}^+$ and $f(0) = 0$, ii) $g < 0$, $g' < 0$ in $\mathbb{R}^+$ and $g(0) = 0$.

We remark here that functions f and g , defined by [We]: $g(x) = -x^{\nu}(= -f(x))$ or by a Langmuir type condition [Do] : $g(x) = -a\ x^n/(b + cx^n)$ $(= -f(x))$, verify conditions (3i,ii) for all constants a, b, c, n, $\nu > 0$.

Firstly, we study an auxiliary moving boundary problem. We generalize the results obtained in [FaPr1,FaPr2] changing the nonlinear condition on the fixed face $x = 0$ by other one on the moving boundary $x = s(t)$, given by (2i). Secondly, we study the Wen—Langmuir free boundary model for noncatalytic gas—solid reactions that consists in finding $T > 0$, $x = s(t)$ and $u = u(x,t)$ such that they satisfy conditions (2). We prove that there exists a unique solution for a sufficiently small $T > 0$. Moreover, the solution is given through the unique fixed point, in an adequate Banach space, of the following contraction operator F_2 : For $s = s(t) \in C^0([0,T])$ we define

(4) $$F_2(s)\,(t) = \int_0^t f(v(s(\tau), \tau))\ d\tau$$

where v is the solution of problem (1i—ii—iv) and (2i). Here we exploit some techniques used in [CoRi] for sorption of swelling solvents in polymers.

V. On the free boundary problem for the Michaelis—Menten absorption model for root growth

Goal : "We give a growth absorption model for the surface of a root of a plant through an absorption mechanism. For low concentrations the resultant equations have been analytically solved by the quasi—stationary method. This solution is used to compute growth of the radius of the root".

194

Many methods exist for studying the mechanism involved in nutrient uptake. One of the most promising methods is the mathematical model, which can be a satisfactory method of modelling the plant—root system by use of the partial differential equation for convective and diffusive flow to a root [Cu1,Cu2]. In general, these models have not considered computing root growth, but rather they have assumed young roots to be growing at exponential rates. We compute the free boundary (the root—soil interface) through the quasi—stationary method. We obtain an analytical solution for the nutrient interface concentration and the interface position. Taking into account the idea of the model used for the shrinking core problem for noncatalytic gas—solid reactions [TaVi], we propose the following free boundary problem for root growth assuming low concentrations (in cylindrical coordinates) [ReTaCa] :

i) $\quad D\,C_{rr} + D\alpha_0\,\dfrac{C_r}{r} = 0,\qquad s(t) < r < R,\ t > 0,\ \alpha_0 = 1+\epsilon,\ \epsilon = v_0 s_0/Db > 0,$

(1) ii) $\quad C(r,0) = \Phi(r)\,,\qquad s_0 \le r \le R\,,\qquad$ iii) $\ s(0) = s_0\,,\ \ 0 < s_0 < R\,,$

iv) $\quad C(R,t) = C_\infty > 0\,,\ \ t > 0,$

v) $\quad Db\,C_r(s(t),t) + v_0\,C(s(t),t) = k\,C(s(t),t) - E = a\,C(s(t),t)\,\dot{s}(t)\,,\ t > 0.$

The solution of the problem is given by :

(2) $$C(r,t) = \beta(t) - \frac{\alpha(t)}{r^\epsilon}\,,\quad s(t) < r < R\,,\ t > 0\,,$$

where:

(3) $\quad \alpha(t) = \left[\dfrac{1}{D\,b}\right]\dfrac{\left[(k - v_0)C_\infty - E\right]}{\dfrac{\epsilon}{s(t)^{1+\epsilon}} + \dfrac{(k - v_0)}{D\,b}\left[\dfrac{1}{s(t)^\epsilon} - \dfrac{1}{R^\epsilon}\right]}\,,\quad \beta(t) = C_\infty + \dfrac{\alpha(t)}{R^\epsilon}\,,$

(4) $\quad \Phi(r) = C_\infty - \dfrac{\left[(k - v_0)\,C_\infty - E\right]}{\dfrac{v_0}{s_0^\epsilon} + (k - v_0)\left[\dfrac{1}{s_0^\epsilon} - \dfrac{1}{R^\epsilon}\right]}\left[\dfrac{1}{r^\epsilon} - \dfrac{1}{R^\epsilon}\right]$

and s(t) is the unique solution of the following Cauchy problem :

(5) $\quad \dot{s}(t) = F(s(t))\,,\ t > 0,\qquad s(0) = s_0 \in (0,R)\,,$

with:

(6) $\quad F(s) = \dfrac{k}{a}\left[1 - \alpha_3\,H(s)\right]\,,\quad H(s) = \dfrac{\left[1 + \alpha_2\,G(s)\right]}{\left[1 + \alpha_1\,G(s)\right]}\,,\quad G(s) = s\left[1 - \left(\dfrac{s}{R}\right)^\epsilon\right],$

(7) $\quad \alpha_1 = \dfrac{E}{v_0\,s_0\,C_\infty} > 0\qquad,\ \alpha_2 = \dfrac{(k - v_0)}{v_0\,s_0} > 0\qquad,\ \alpha_3 = \dfrac{E}{k\,C_\infty} > 0\,.$

The solution of the problem (5) is computed numerically and the results are plotted for the interface concentration C(s(t),t) vs. s and the interface position s(t) vs. t respectively as a function of the dimensionless parameter k/v_0 [ReTaCa] .

REFERENCES

[CoRi] E. COMPARINI − R. RICCI, "On the swelling of a glassy polymer in contact with a well-stirred solvent", Math. Meth. Appl. Sci., 7(1985), 238−250.

[Cu1] J.H. CUSHMANN, "Analytical study of the effect of ion depletion (replenishment) caused by microbial activity near a root", Soil Science, 129, 2(1980), 69−87.

[Cu2] J.H. CUSHMANN, "Nutrient transport inside and outside the root rhizosphere theory", Soil Science Society of America J., 46,4(1982), 704−709.

[Do] D.D. DO, "On the validity of the shrinking core model in non-catalytic gas solid reaction", Chem. Eng. Sci., 37(1982), 1477−1481.

[FaPr1] A. FASANO − M. PRIMICERIO, "Esistenza e unicità della soluzione per una classe di problemi di diffusione con condizioni al contorno non lineari", Bollettino Un. Matematica Italiana, 3(1970), 660−667.

[FaPr2] A. FASANO −M. PRIMICERIO, "Su un problema unidimensionale di diffusione in un mezzo a contorno mobile con condizioni limiti non lineari", Ann. Mat. Pura Appl., 93(1972), 333−357.

[Go] T.R. GOODMAN, "The heat-balance integral and its applications to problems involving a change of phase", Trans. of the ASME, 80(1958), 335−342.

[MeTa] J.L. MENALDI − D.A. TARZIA, "Generalized Lamé−Clapeyron solution for a one−phase source Stefan problem", Submitted to Mat. Aplic. Comp.

[ReTaCa] J.C.REGINATO− D.A.TARZIA− A.CANTERO, "On the free boundary problem for the Michaelis−Menten absorption model for root growth", Soil Science, 150 (1990), 722−729.

[RiTa] R. RICCI − D.A. TARZIA, "Asymptotic behavior of the solutions of a class of diffusion−reaction equations", in Free Boundary Problems :Theory and Applications, 11−20 June 1987, Irsee/Bavaria, K.H. Hoffmann−J. Sprekels (Eds.), Research Notes in Math. No 185, Longman, Essex (1990), 719−721.

[SoWiAl] A. D. SOLOMON − D. G. WILSON − V. ALEXIADES, "A mushy zone model with an exact solution", Letters Heat Mass Transfer, 9 (1982), 319−324.

[St] I. STAKGOLD, "Reaction-diffusion problems in chemical engineering", A. Fasano−M. Primicerio (Eds.), Lecture Notes in Math. N^0 1224, Springer Verlag (1986), 119−152.

[Ta1] D.A. TARZIA, "Determination of unknown thermal coefficients of a semi-infinite material for the one-phase Lamé-Clapeyron (Stefan) problem through the Solomon-Wilson-Alexiades mushy zone model", Int. Comm. Heat Mass Transfer, 14 (1987), 219 − 228.

[Ta2] D.A. TARZIA, "A bibliography on moving−free boundary problems for the heat−diffusion equation. The Stefan Problem", Firenze (1988) with 2528 references.

[Ta3] D.A. TARZIA, "A variant of the heat balance integral method and a new proof of the exponentially fast asymptotic behavior of the solutions in heat conduction problems with absorption", Int. J. Engineering Science, 28 (1990), 1253−1259.

[Ta4} D.A. TARZIA, "Comportamiento asintótico exponencial en la ecuación de medios porosos con absorción", Cuadernos Inst. Matemática "B. Levi", Rosario, 17(1989), 73−86.

[Ta5] D.A. TARZIA, "Neumann−like solution for the two−phase Stefan problem with a simple mushy zone model", Mat. Aplic. Comp., 9 (1990), 201−211.

[TaVi] D.A. TARZIA − L.T. VILLA, "On the free boundary problem in the Wen-Langmuir shrinking core model for noncatalytic gas-solid reactions", Meccanica, 24(1989), 86−92.

[We] C.Y. WEN, "Noncatalytic heterogeneous solid fluid reaction models", Industrial Eng. Chem., 60 No. 9(1968), 33−54.

PROMAR (CONICET−UNR), Instituto de Matemática "Beppo Levi",
Fac. Ciencias Ex. Ing. Agr., Univ. Nac. de Rosario,
Av. Pellegrini 250, 2000 Rosario, Argentina.

J VINALS

Dynamic scaling during interfacial growth in the one-sided model of solidification

The one-sided model was originally introduced as a model for solidification from the melt. There, solute diffusion coefficients in the solid phases are normally much smaller than in the fluid phases. In some cases (e.g., nearly stoichiometric compounds), it is justified to neglect diffusion in the solid phase altogether. The model that we consider here is somewhat simpler than the one-sided model used in solidification studies, but the essential feature common to both is the neglect of changes in the order parameter in one of the phases.

We thus consider a system which is infinite in extent and that, in equilibrium, two phases coexist across a planar interface which is taken to be the plane $z = 0$. The phase with non-zero diffusion coefficient (or α-phase) occupies the space $z \geq 0$, whereas the rest of the system is occupied by the other phase. Macroscopic order parameter inhomogeneities in the α-phase relax diffusively according to

$$\frac{\partial \phi}{\partial t} = D\nabla^2\phi, \tag{1}$$

where $\phi(x, z, t)$ is the macroscopic order parameter and D is the diffusion coefficient, assumed to be constant. We also introduce the quasistationary approximation in which retardation effects are neglected, and replace the full diffusion equation by Laplace's equation,

$$\nabla^2\phi = 0. \tag{2}$$

This approximation is valid when the time scale of motion of the interface is much slower than the relaxation times for order parameter inhomogeneities in the bulk phases.

In order to specify the model completely, this equation has to be supplemented with appropriate boundary conditions at the interface. Conservation of order parameter leads to

$$- D\nabla\phi|_{Int} \cdot \hat{n} = (\Delta\phi)\, v_n, \tag{3}$$

where the unit normal $\hat{n}$ is directed into the α-phase, v_n is the local normal velocity of the interface and $\Delta\phi$ is the equilibrium miscibility gap. The subscript Int indicates that the quantity is to be evaluated as the limit approaching the interface from the α phase. We also assume local equilibrium at the interface, such that the order parameter satisfies a Gibbs-Thomson relation of the form,

$$\phi|_{Int} = \phi^{eq} + \Gamma\kappa, \tag{4}$$

where ϕ^{eq} is the equilibrium value of the order parameter of the $\alpha-$ phase at co-existence across a planar interface. κ is the mean curvature of the interface, taken as positive if the nearest center of curvature lies in the α phase. The coefficient Γ contains thermodynamic in formation and is of the form $\Gamma = \sigma\chi/\Delta\phi$, where σ is the excess surface free energy, and χ is the order parameter susceptibility. We will allow an explicit dependence of Γ on the orientation of the interface. Under these conditions, the coefficient Γ appearing in Eq. (4) is of the form,

$$\Gamma = \gamma(\theta) + \frac{\partial^2\gamma(\theta)}{\partial\theta^2}. \tag{5}$$

where $\theta = \hat{n}\cdot\hat{k}$ is the orientation of the local normal relative to the z-axis, and $\gamma(\theta)$ is the excess surface free energy which now depends explicitly on orientation.

We use in this study a constant flux boundary condition far from the interface: $\vec{j}(z = +\infty) = j_0\hat{k}$ ($j_0 > 0$). Eq.(2) together with the boundary conditions (3)-(4) admits a steady state in which a planar interface advances at constant velocity, $v = j_0/\Delta\phi$. Furthermore, the order parameter in the α phase is given in this steady state by

$$\phi^{(0)}(z) = \phi^{eq} - \frac{j_0}{D}\left(z - \frac{j_0 t}{\Delta\phi}\right), \quad z \geq \frac{j_0 t}{\Delta\phi}. \tag{6}$$

We introduce dimensionless variables in the following way. First we define a dimensionless order parameter $u(x, z, t)$,

$$\phi(x, z, t) = \phi^{eq} - \frac{j_0}{D}\left(z - \frac{j_0 t}{\Delta\phi}\right) + \Delta\phi u(x, z, t). \tag{7}$$

We also rescale lengths by a length scale, to be specified momentarily, $\mathcal{L}$, and times by $\mathcal{L}^2/D$. We define $1/l = j_0/\Delta\phi D\mathcal{L}$, and $d_0 = \mathcal{L}\Gamma/\Delta\phi$. We also introduce

$\xi(x,t) = h(x,t) - vt$, which is the interface displacement relative to a uniformly advancing front with velocity v.

The partial differential equation for $u(x,z,t)$ and the boundary conditions prescribed on the moving interface can be recast into an equation of motion for the interface alone. We consider a system infinite in the z-direction, and of finite width, W, in the x-direction satisfying periodic boundary conditions. We define the Green's function: $\nabla^2 G(\vec{r},\vec{r'}) = -\delta(\vec{r} - \vec{r'})$, also satisfying periodic boundary conditions in the x-direction. The explicit form of the Green's function reads

$$G(\vec{r},\vec{r'}) = -\frac{1}{2W}|\Delta z| - \frac{1}{4\pi}ln\left[1 - 2pcos\left(q_0\Delta x\right) + p^2\right], \tag{8}$$

where $\vec{r} = (x,z)$, $\Delta z = z - z'$, $\Delta x = x - x'$, $q_0 = 2\pi/W$, and $p = e^{-q_0|\Delta z|}$. With this choice of Green's function and the boundary conditions given above, the local normal velocities satisfy

$$\frac{1}{2}\left(d_0\kappa(s) + \frac{\xi(s)}{l}\right) - \int_0^{S_{tot}} ds'\hat{n}\cdot\nabla'G(s,s') \times$$

$$\left(d_0\kappa(s') + \frac{\xi(s')}{l}\right) = \int_0^{S_{tot}} ds'G(s,s')v_n(s'), \tag{9}$$

where the interface has been parametrized by a contour variable s. s and s' denote two arbitrary points on the interface and the integration extends over the entire interface.

Equation (9) is our starting point to study the evolution of the interface. It is known from linear stability theory that the steady state Eq. (6) corresponding to an advancing planar front is unstable against infinitesimal perturbations of wavenumber $q < q_c = (d_0l)^{-1/2}$. We analyze in the present work the growth of the instability away from a planar interface well beyond the linear regime.

We describe the interface by a discrete set of points and numerically solve for its motion in the following way: at a given time t the interface configuration is specified, and, thus, the left-hand side of Eq. (9) is known. Equation (9) is an integral equation for the normal velocities v_n. Once the normal velocities have been determined, the new interface position is calculated by forward integration in time.

Equation (9) determines the normal velocities only; we have found it useful to add a tangential component, v_t, to keep the interface nodes equally spaced in time. We rewrite the equation of motion for the discrete interface as a system of ordinary differential equations (ODE) for θ and the total arc length S_{tot},

$$\frac{\partial\theta}{\partial t} = v_t\kappa + \frac{\partial v_n}{\partial s}, \tag{10}$$

199

$$\frac{dS_{tot}}{dt} = -\int_0^{S_{tot}} \kappa(s)v_n(s)ds, \tag{11}$$

where the tangential component of the velocity is determined by,

$$v_t(s) - v_t(s = 0) = sg(S_{tot}) - S_{tot}g(s), \tag{12}$$

with,

$$g(s) = -\frac{1}{S_{tot}} \int_0^s v_n(s')\kappa(s')ds'. \tag{13}$$

In summary, if the interface is represented by a set of N equally spaced points along the contour, Eqs. (10) and (11) constitute a system of ordinary differential equations, once the velocities are known. We integrate this system of equations with an implicit ODE solver (IVPAG on an ETA-10G or HSODEN on a Cray Y-MP). The value of N is dynamically determined during the numerical solution. Additional nodes are added such that $\kappa_{max}S_{tot}/N < 1$ at all times, where κ_{max} is the maximum curvature of the interface. A typical run starts with $N = 256$. N is incremented in steps of 96 when required by the above condition. At the latest times studied, N reaches $N = 736$.

The choice of system size and values of the parameters defining the model is as follows: we take d_0 of the order of unity, as the smallest (dimensionless) length. We have also chosen $l = 40$ and $W = 800$. In real systems l can be perhaps five or six orders of magnitude larger than d_0 and, ideally, $W \gg q_c^{-1} \sim \lambda_c$, the longest wavelength for instability in linear theory. The values that we have chosen represent a compromise between computational constraints and the need to reach length scales over which the nonlinear evolution of the interface, free of finite-size effects, can be studied. We have studied both isotropic and anisotropic interfacial boundary conditions. Specifically, we have taken

$$d_0 = \overline{d}_0 \left(1 - \epsilon cos4\theta\right), \tag{14}$$

and have considered parameter values of $\overline{d}_0 = 1, 1/4$, and $1/16$, and $\epsilon = 0$ and 0.1.

The initial condition is taken to be a linear combination in q-space of the modes that are linearly unstable. Higher modes do not seem to modify our results but cause numerical difficulties. The complex coefficients in the linear combination are chosen at random according to a Gaussian distribution with zero mean and width equal to one. With this choice, the maximum initial amplitude of any mode is about

2% of its wavelength. The results have been averaged over an ensemble of initial conditions, typically from fifty to one hundred in the one-sided model, and around twenty for the results that will be shown for the two-sided symmetric model.

To characterize the growth of the pattern quantitatively we have defined two independent measures of its characteristic length: i) the mixing zone, ΔZ, defined as the maximum peak-to-peak distance in the pattern, and, ii) the root-mean-square displacement of the interface from planarity, z_2. It is shown that both measures of the characteristic length are proportional after an initial transient. Such a proportionality provides evidence that there is only one characteristic length scale for the spatial pattern. We further show that asymptotically either length becomes proportional to time.

Additional insight into the the detailed structure of the pattern can be gained from the power spectrum of the interface,

$$P(q,t) = \frac{1}{N^2} |\sum_{i=1}^{N} \xi(s_i) e^{iqs_i}|^2. \tag{15}$$

It is shown that the power spectrum satisifies a generalized scaling relation,

$$P(q,t) = \Delta Z(t)^2 \mathcal{F}(q\Delta Z(t)) \tag{16}$$

where $\mathcal{F}(x)$ is independent of time.

We further develop a renormalization procedure, similar in spirit to Monte Carlo renormalization group, for the type of interface equations being considered. Evidence has been presented for the existence of a scaling regime in the growth process. The renormalization procedure both exposes the scaling and provides a means for evaluating scaling exponents.

We find that the recursion relation for a characteristic length scale R is

$$R(t; l, d_0) = bR(b^{\lambda_t}t; b^{1+\lambda_t}l, b^{-(3+\lambda_t)}d_0), \tag{17}$$

where b is the length rescaling factor and λ_t an expoenent to be determined. Formally by choosing $b^{\lambda_t}t = 1$, one obtains,

$$R(t; l, d_0) = t^{-1/\lambda_t}R(1; t^{-(1+\lambda_t)/\lambda_t}l, t^{(3+\lambda_t)/\lambda_t}d_0). \tag{18}$$

Pure power law growth (at sufficiently late time) will occur if a fixed point of the renormalization transformation exists. One can see from Eq. (17) the appearance of two types of fixed points. First, a fixed point will exist for any d_0 if $\lambda_t = -3$. Note in this case l iterates rapidly to zero under the renormalization transformation. This fixed point describes a system in which the driving flux is arbitrarily large. The continuum description and the interface equations are questionable in this limit. The other (physically meaningful) fixed point behavior occurs for $\lambda_t = -1$. This fixed point exists for all $l > 0$, and is characterized by the strong "irrelevance" of the stabilizing term d_0 which then decreases as $d_0' = d_0/b^2$. This choice of $\lambda_t = -1$, corresponds to the growth of a characteristic length as $R \sim t$, which has been observed numerically.

This analysis indicates that the term proportional to the curvature in the interface equation (stabilizing contribution) is not important as short length scale structures are eliminated by the renormalization procedure, and hence, contrary to the situation with regard to steady states and their stability, the introduction of anisotropy in the surface tension does not seem to play an important role in the scaling (since it appears in a strongly irrelevant term). This is an interesting prospect, one that will be examined further within the context of the dynamical evolution in anisotropic Hele-Shaw cells. Finally, the scaling argument suggests a scheme, analogous to Monte Carlo renormalization group for numerically effecting the renormalization transformation. Such a procedure will also be discussed in detail.

The results obtained are consistent with the temporal laws discussed above and will be presented. Systems with different driving fluxes and anisotropic boundary conditions have been studied as well. We conclude that there exists a universal form for the scaled power spectrum, independent of the externally imposed flux (which is proportional to $1/l$), and of the anisotropy of the boundary conditions. A similar analysis for the symmetric model leads to the same conclusions.

Supercomputer Computations Research Institute, B-186
Florida State University
Tallahassee, Florida 32306-4052

J-J XU

Interfacial wave theory of solidification – dendritic pattern formation and selection of tip-velocity

INTRODUCTION

Dendritic growth is a common phenomenon in phase transition and crystal growth. The experimental observations show that at the later stage of growth, a dendrite has a smooth tip moving with a constant velocity; meanwhile it emits a stationary wave-train, propagating along the interface towards the root. The essence and origin of this non-linear interfacial phenomenon have been a fundamental subject in the field of condensed matter physics and material science for a long period of time [1-21]. The understanding to this problem has a great significance to a much broad area, such as fluid-dynamics, chemical engineering, biological science, etc., where similar pattern formation phenomena occur.

In the past two decades, around this subject there has been proposed a number of theories. One of the most popular theories is so-called "Microscopic Solvability Condition (MSC) theory", posed by several groups of investigators: Langer, Levine and Kessler, Pomeau and Pelce, Bensimon et. al., etc. ([8-14]). The MSC theory considers the steady state solution separately from the unsteady dendrite growth. The MSC-theory states that: (1) for the isotropic surface tension case, the system does not allow a steady, smooth, needle-like solution, which is smooth at the tip, while approaches the Ivantsov solution at far field (satisfying so-called Nash-Glicksman condition); (2) with the inclusion of small amount of anisotrophy of surface tension, the system then permits a discrete set of steady, smooth, needle-like solutions. These results were first drawn from a local model problem by a non-rigorous way. Most recently, the proof has been refined by some investigators (refer to [22-24]). The above two conclusions are now more likely to be correct even for the original, fully non-local problem. Nevertheless, the MSC-theory does not justify that why, from the physical point of view, the Nash-Glicksman's far field condition is necessary for the actually finite growth system. For a more detailed discussion on this issue, readers are referred to [17]. Furthermore, in order for the MSC-theory to solve the selection problem, an additional stability argument was posed; that is (3) among the above steady needle-like solutions only the solution with the largest growth velocity is stable; all the other solutions are unstable. It is just upon this stability argument, the MSC-theory claimed that the unique stable, steady solution is selected. Obviously, according to the MSC theory, the inclusion of the anisotropy of surface tension is a necessary condition for the dendritic growth.

The argument (3) about the stability of solution was first made by Kessler and Levine in 1986 through the numerical experiments [10]. This stability argument was claimed to be also proven analytically by Bensimon et. al. in 1987 ([11]).

In this talk, I intend to present some analytical results, which disagree with the above MSC theory. In the recent years, a series of investigations concerned with the global instability mechanisms of solidification systems has been carried out, which results in a wave theory to resolve the long-standing problems — the pattern formation and the selection of tip-velocity ([17-21]). One of the most important results drawn from these investigations is that the selection condition of the dendrite's tip-velocity can be found even in the absence of the anisotropy of surface tension. we call this new theory "Interfacial Wave Theory". The major conclusions of the present theory are:

1. The dendrite growth is intrinsically a time-dependent wave phenomenon. For the case of non-zero surface tension, introducing the steady state of needle-like crystal growth to the problem, physically, is not meaningful; and mathematically, is also unnecessary. As the surface tension is not zero, the actually selected solution for a realistic dendrite growth is not the stable, steady, needle-like, neighbouring solution of the Ivantsov solution, it is the time-dependent, Global, Neutrally-Stable, neighbouring solution of the Ivantsov solution.

2. The anisotropy of surface tension is not a necessary condition for dendrite growth. Its effect on the selection condition of tip-velocity, for the simplicity, can be neglected as other secondary effects.

3. The Nach-Glicksman's far field condition should be modified (see [17]). For the unsteady dendrite growth one needs to apply the radiation condition in the far field.

4. In the dendritic growth system there exists a special simple turning point ξ_c, which plays a vital role for the pattern formation and the selection of tip velocity. The location of this turning point is related to the eigenvalue of the unsteady solution. This special turning point was missed in the analysis by Bensimon, Pelce and Shraiman in 1987 ([11]), as well as in the analytical works by other authors of MSC theory. It was just for this reason that Bensimon and other authors were unable to explore the global instability mechanism existing in the system. The existence and significance of this turning point ξ_c was first identified by J.J. Xu in 1989 (see [18]).

5. The global instability mechanism of the dendrite growth system is an entirely new instability mechanism generated by the above turning point. It is determined by the wave interactions at the turning point and the leading edge of the tip. The global instability mechanism of the system is explored in the "Interfacial Wave Theory", it is called the Global Trapped Wave (GTW) instability mechanism.

6. The stability parameter $\varepsilon = \frac{\Gamma^{1/2}}{\eta_0^2}$ is the only control parameter that the system needs, where $\Gamma = \ell_c/\ell_T$, ℓ_c is the capillary length, ℓ_T is the thermal diffusion length and η_0^2 is a function of the undercooling temperature. It is found that for any given ε, the system permits a discrete set of unstable, Global Trapped Wave (GTW) modes; furthermore, the system has a unique Global Neutral Stable (GNS) mode, as $\varepsilon = \varepsilon_* = 0.1470$. The interface shape of dendrite in the GNS state for the case $T_\infty = -0.06844$ is shown in Figure 1.

7. The above Global Neutral Stability condition: $\varepsilon = \varepsilon_* = 0.1470$, is the selection condition of the realistic solution, which yields the tip velocity of dendrite at the final stage of growth. For more complicated systems, the value of ε_* may be a function of other physical parameters, such as the anisotropy of surface tension, the difference of densities between two phases, etc.. It has been examined, however, that the effects of these parameters are insignificant. The present theory is in a reasonably good agreement with the experimental observations.

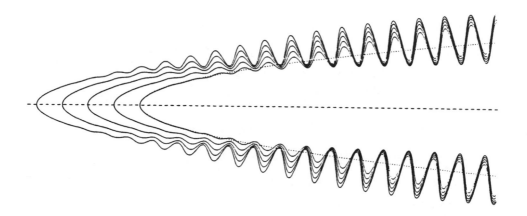

Figure 1. Interface shape of the growing dendrite in GST state.

REFERENCES

1. Ivantsov, G. P., Dokl. Akad. Nauk, SSSR. **58**, p. 567, (1947).

2. Horvay, G. and Cahn, J. W., Acta Metall. **9**, p. 695, (1961).

3. Langer, J. S., Rev. Mod. Phys. Vol. **52**, No: 1, January (1980).

4. Nash, G. E. and Glicksman, M. E., Acta Metall. **22**, p. 1283, (1974).

5. Langer, J. S. and Müller-Krumbaar, Acta Metall., **26**, p. 1681; 1689; 1697; (1978).

6. Glicksman, M. E., Schaefer, R. J. and Ayers, J. D., Metall. Trans., A 7, p. 1747, (1976).

7. Glicksman, M. E., Schaefer, R. J. and Ayers, J. D., Phil. Mag., **32**, p. 725, (1975).

8. Langer, J. S., *Lectures in the Theory of Pattern Formation* USMG NATO AS Les Houches Session XLVI 1986 - Le hasard et la matiere/ chance and matter. Edited by J. Souletie, J. Vannimenus and R. Stora. Elsevier Science Publishers

9. Kessler, D. A., Koplik, J. and Levine, H., *Pattern Formation Far From Equilibrium: the free space dendritic crystal* in "Proc. NATO A.R.W. on Patterns, Defects and Microstructures in Non-equilibrium Systems" (Astin, TX., March, 1986)

10. Kessler, D. A. and Levine, H., Phys. Rev. Lett., 57, p. 3069, (1986).

11. Bensimon, D., Pelce, P. and Shraiman, B. I., J. Physique 48, p. 2081, (1987).

12. Pelce, P. and Pomeau, Y., Studies in Applied Mathematics **74**, p. 245, (1986).

13. Barber, M. N., Barbieri, A. and Langer, J. S., The Physical Rev. Vol. 36, No. 7, p. 3340, (1987).

14. Langer, J. S., The Physical Rev. A. Vol. 36, No. 7, p. 3350, (1987).

15. Xu, J. J., *Global Asymptotic Solution for Axisymmetric Dendrite Growth with Small Undercooling* in "Structure and Dynamics of Partially Solidified System"; p. 97. Edited by D.E. Loper NATO ASI Series E. No: 125 (1987).

16. Xu, J. J., The Phys. Rev. A. Vol. 37, No: 8, p. 3087, (1988).

17. Xu, J. J., Studies in Applied Mathematics 82: p. 71-91, (1990).

18. Xu, J. J., *Interfacial Wave Theory for Dendritic Structure of a Growing Needle Crystal (I). Local Instability Mechanism.*
The Phys. Rev. A., 40, No: 3, p. 1599, (1989).
Xu, J. J., *Interfacial Wave Theory for Dendritic Structure of a Growing Needle Crystal (II). Wave Emission Mechanism at Turning Point.*
The Phys. Rev. A., 40, No: 3, p. 1609, (1989).

19. Xu, J. J., *Interfacial Wave Theory For Dendritic Growth — Global Mode Solutions and Quantum Condition*   Canadian J. of Physics, (January 1990)

20. Xu, J. J., *Global Wave Mode Theory For The Formation of Dendritic Structure On A Growing Needle Crystal*
Physica Status Solidi **b**, (February 1990)

21. Xu, J. J., *Global Neutral Stability of Dendrite Growth and Selection Condition of Tip-Velocity*
J. Crystal Growth, (In press, 1990)

22. Kruskal, M. and Segur, H., *Asymptotics Beyond All Orders*
(Unpublished paper, 1985.)

23. Amick, C., and Mcleod, B., (Private communication, 1989.)

24. Hammersley, J. M., Mazzarino, G., IMA Journal of Applied Mathematics **42**, p. 43. (1989).

JIAN-JUN XU
Department of Mathematics and Statistics
McGill University
Montreal, Quebec, Canada H3A 2K6

S ZHENG
Some results on phase field equations

We would like to present in this paper some recent results on the phase field equations. A part of them was obtained with C. Elliott and J. Yong.

The following phase field equations

$$\tau \phi_t = \xi^2 \Delta \phi + \frac{1}{2}(\phi - \phi^3) + 2u$$

$$u_t + \frac{\ell}{2}\phi_t = K \Delta u$$

(1)

were proposed by Caginalp [1] to study the phase transitions with finite thickness in solidification. In (1) the unknown functions ϕ and u represent the phase function and the reduced temperature, respectively. The positive parameters τ, ξ, ℓ, K represent the relaxtion time, a length scale, which, at the microscopic level, is a measure of the strength of the bounding, the latent heat and the thermal diffusivity, respectively.

The global existence of solutions, among other things, to the Dirichlet initial boundary value problem for the equations (1) has been proved by Caginalp in [1] under the restriction on the coefficients: $\xi^2/\tau < K$. In a recent paper [2] Elliott and Zheng Songmu improved these results using a different method. The global existence of solutions to both Dirichlet and Neumann initial boundary value problems has been proved without the restriction on the coefficients. The asymptotic behavior of solutions as $t \to +\infty$ and the corresponding stationary problems are also extensively studied. In particular, it has been proved that as time goes to infinity, the temperature u always converges to the equilibrium in the case of Dirichlet initial boundary value problem. More precisely, in the case of Dirichlet initial boundary value problem we have

Theorem 1.   For any initial data $\phi_0, u_0 \in H^2(\Omega)$ and boundary data $\phi_\Gamma(x), u_\Gamma(x) \in H^{3/2}(\Gamma)$ satisfying the compatibility conditions, the Dirichlet initial boundary value problem for the equations (1) admits a unique global smooth solution $(\phi(x,t), u(x,t))$. Moreover, the ω - limit set $\omega(\phi_0, u_0)$ is a connected compact (weakly in H^2, strongly in H^1) subset which consists of the equilibria. In particular, we have

$$\lim_{t \to \infty} u(x,t) = \bar{u}(x), \;\; weakly \; in \; H^2, \; strongly \; in \; H^1$$

(2)

The proof of the theorem based on the following observation:

$$V(t) = \int_\Omega \left(\frac{\xi^2}{2}|\Delta\phi|^2 + \frac{1}{2}(\frac{1}{4}\phi^4 - \frac{1}{2}\phi^2) + 2v^2 + 2\phi\bar{u}\right)dx \tag{3}$$

with $v = u - \bar{u}$, serves as the Liapunov functional for the problem. In the above, $\bar{u}$ is the equilibrium, i.e,

$$\Delta u = 0, \ in\Omega$$
$$u|_\Gamma = u_\Gamma(x) \tag{4}$$

Thus the a priori estimates of H^1 norm of ϕ and L^2 norm of v follow. By energy method we obtain other desired estimates.

In a recent paper [3] Penrose and Fife proposed thermodynamically consistent models of phase-field type for the kinetics of phase transitions. Raised on their theory, they suggested to consider the following equations

$$\phi_t = K_1\left(\frac{2a}{Y_0}(1 - \frac{u}{T_0}) + s'(\phi) + \kappa_1\Delta\phi\right) \tag{5}$$

$$u_t - 2a\phi\phi_t = K_2\Delta u \tag{6}$$

instead of (1). In the above K_1, K_2, T_0, a and κ_1 are constants.

Constructing the corresponding Liapunov functional and making the minor changes in the energy estimates, we have proved in [6] the following

Theorem 2. For any initial data $\phi_0, u_0 \in H^2(\Omega)$ and boundary data $\phi_\Gamma(x), u_\Gamma(x) \in H^{3/2}(\Gamma)$ satisfying the compatibility conditions, the Dirichlet initial boundary value problem for the equations (5) and (6) admits a unique global smooth solution $(\phi(x,t), u(x,t))$. Moreover, the $\omega-$limit set $\omega(\phi_0, u_0)$ is a connected compact (weakly in H^2, strongly in H^1) subset which consists of the equilibria. In particular, (2) holds for the temperature u.

Similar conclusions hold for the Neumann initial boundary value problem.

As pointed out above, we have proved in [2] that the stationary problem has in general many solutions. Thus an interesting control problem arises: Whether we can construct a feedback control so that the states, in particular the phase function ϕ, will converge to a prescribed equilibrium? More precisely, we accordingly consider the following equations:

$$\tau\phi_t = \xi^2\Delta\phi + \frac{1}{2}(\phi - \phi^3) + 2u \tag{7}$$

$$u_t + \frac{\ell}{2}\phi_t = K\Delta u + f \tag{8}$$

with f being a feedback control variable.

By explicitly constructing the feedback control variable f we have proved in [5] the following

Theorem 3. For any initial data ϕ_0, u_0 in H^2 and prescribed equilibrium $\bar{\phi}$,we can construct an explicit feedback control f so that as time goes to infinity the phase function $\phi(x,t)$ will converge to the prescribed equilibrium $\bar{\phi}(x)$.

Remark. As stated in Theorem 1, the temperature $u(x,t)$ always converges to the equilibrium $\bar{u}$.

The corresponding feedback stabilization results for the Cahn-Hilliard equation have already been obtained in [4].

References
[1] G. Caginalp, An analysis of a phase field model of a free boundary, Arch. Rat. Mech. Anal., **92** (1906), 205-245.
[2] C. Elliott and Zheng Songmu, Global existence and stability of solutions to the phase field equations, Preprint No.74, Bonn University, SFB 256, 1989.
[3] O. Penrose and P. C. Fife, Thermodynamically consistent models of phase-field type for the kinetics of phase transitions, preprint, 1989.
[4] J. Yong and S. Zheng, Feedback stabilization and optimal control for the Cahn-Hilliard equation, preprint, 1989.
[5] J. Yong and S. Zheng, Feedback stabilization for the phase field equation, preprint, 1990.
[6] S. Zheng, Global existence of solutions to the modified phase field equations, preprint 1990.

Zheng Songmu
Institute of Mathematics
Fudan University
Shanghai 200433, China.

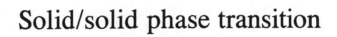

Solid/solid phase transition

H W ALT AND I PAWLOW

Solutions for a model of non-isothermal phase separation

We consider a mathematical model of non-isothermal phase separation in a two-component system. It consists of the fourth order Cahn-Hilliard equation for the mass concentration c and a second order equation for the Kelvin temperature θ. In terms of the variables

$$u = c, \quad v = \frac{\mu}{\theta}, \quad w = \frac{1}{\theta} > 0 \quad ,$$

where μ is the difference between the chemical potentials of both components, the governing equations are

$$(1) \qquad\qquad -v + \varphi_{,u}(u, w) - \nabla \cdot (\kappa w \nabla u) = 0 \quad ,$$

$$(2) \qquad\qquad \partial_t u - \nabla \cdot (l_{11}(u, v, w)\nabla v - l_{12}(u, v, w)\nabla w) = 0 \quad ,$$

$$(3) \qquad -\partial_t E(u, w) - \nabla \cdot (l_{22}(u, v, w)\nabla w - l_{21}(u, v, w)\nabla v) + g = 0$$

in a time-space cylinder $\Omega_{t_0} :=]0, t_0[\times \Omega$, where $\Omega \subset \mathbb{R}^n, n \geq 1$, is a domain with Lipschitz boundary Γ, and where κ is a positive constant.

In the isothermal case w=const., equations (1), (2) originally have been introduced by Cahn and Hilliard [4, 5]. They describe what happens if the system subject to an instantaneous quench is brought into a state which is not thermodynamically stable. The Cahn-Hilliard model neglects all thermal effects and focuses only on mass diffusion.

Mathematical and numerical aspects of the Cahn-Hilliard equation were extensively studied, see [6 − 12].

Since in realistic physical systems quenches are carried out during finite periods of time, the coupling between the diffusive and thermal phenomena is to be accounted, especially in systems with rapid diffusion time scale. Besides, external thermal activation itself can be used to control the phase separation dynamics.

We have proposed the above extension of the Cahn-Hilliard model to the non-isothermal case, see [2]. The model is based on the Landau-Ginzburg free energy functional

$$(4) \qquad \Phi_\Omega(u, w) = \int_\Omega \Phi(u, w)dx$$

where the energy density Φ is given by

$$(5) \qquad \Phi(u, w) = \varphi(u, w) + \frac{1}{2}\kappa w|\nabla u|^2$$

for functions u and w. It comprises the volumetric energy density $\varphi : \mathbb{R} \times \mathbb{R}_+ \to \mathbb{R}$ of a homogeneous system and the gradient term with a positive constant κ, which represents an interfacial energy between the phases.

System (1)-(3) is derived within the frame of non-equilibrium thermodynamics. Its underlying components are as follows. For the reduced chemical potential v we set

$$(6) \qquad v := \frac{\delta}{\delta u}\Phi(u, w) = \varphi_{,u}(u, w) - \nabla \cdot (\kappa w \nabla u)$$

which is equation (1).

The internal energy density E of the system and the entropy density S are given by Gibbs relation

$$E = \Phi_{,w} \quad \text{and} \quad S = w\Phi_{,w} - \Phi \quad .$$

According to (5) we define the volumetric parts of E and S by

$$e := \varphi_{,w} \quad \text{and} \quad s := w\varphi_{,w} - \varphi \quad .$$

Therefore, since κ is constant,

$$(7) \qquad E(u, w) = e(u, w) + \frac{\kappa}{2}|\nabla u|^2 \quad \text{and} \quad S(u, w) = s(u, w) \quad .$$

The spatial distribution of concentration and temperature is governed by the balance equations

$$\partial_t u + \nabla \cdot \vec{j} = 0, \quad \partial_t E + \nabla \cdot \vec{q} = g \quad ,$$

wher $\vec{j}$ is the mass flux of the selected component and $\vec{q}$ the energy flux. g is a given heat source density. As constitutive relations for these fluxes we use

(8)
$$\vec{j} = -l_{11}\nabla v + l_{12}\nabla w, \quad \vec{q} = l_{22}\nabla w - l_{21}\nabla v \quad ,$$

where $(l_{ij})_{ij}$ is strictly positive definite with bounded entries $l_{ij} = l_{ij}(u, v, w)$. This leads to the differential equations (2) and (3).

The initial state of the parabolic system (1)-(3) is given by

(9)
$$u(0, x) = u_0(x) \quad \text{and} \quad w(0, x) = w_0(x) \quad \text{for} \quad x \in \Omega \quad .$$

On the boundary $\Gamma_{t_0} =]0, t_0[\times\Gamma$ we assume the conditions

(10)
$$\nabla u \cdot \vec{n} = 0, \quad \vec{j} \cdot \vec{n} = 0, \quad \vec{q} \cdot \vec{n} + p(u, v, w) = 0 \quad ,$$

where $\vec{n}$ denotes the outward normal to $\Gamma = \partial\Omega$. The first condition in (10) is natural for the functional (4). The second one expresses mass isolation and the third refers to a heat exchange through the boundary.

The main feature of system (1)-(3) is related to the different qualitative behavior of the free energy density φ as a function of u for different temperatures. At high temperatures φ is convex in u, whereas below a critical temperature θ_{crit} it assumes a characteristic double-well form. As a function of the inverse temperature φ is assumed to be concave. Therefore the main structural assumptions on the free energy read as follows: φ admits a splitting $\varphi = \varphi^0 + \varphi^1$ with C^2-functions φ^0 and φ^1 satisfying

(11)
$$-\varphi_{,ww}(u, w) > 0, \quad \varphi^0_{,uu}(u, w) \geq 0 \quad ,$$

$$\pm\varphi^0_{,u}(u, w) \to \infty \quad \text{for} \quad \pm u \to \infty \quad .$$

Besides, growth conditions on φ are imposed (see [3]). In particular, $\varphi^0_{,u}$ and $\varphi^1_{,u}$ are assumed to have linear growth with respect to w. This is consistent with the concavity condition in (11) which implies that φ in its positive part is sublinear in w. We remark that no assumption on the growth of φ^0 as a function of u is required. We note also that according to the Gibbs relation the specific heat coefficient C_V satisfies

$$C_V := E_{,\theta} = -w^2\varphi_{,ww} > 0 \quad .$$

Further structural assumptions concern the internal energy e and the entropy s. The parabolic norm of the system will be given through the inequality

(12)
$$\alpha_0 e(u, w) - s(u, w) \geq c(\psi_0(u) + \psi_1(w)) - C$$

216

for some positive constants α_0, c and C, where ψ_0 and ψ_1 are non-negative with

$$\frac{\psi_0(u)}{|u|} \to \infty \quad \text{as} \quad |u| \to \infty, \quad \psi_1(w) \to \infty \quad \text{as} \quad w \to 0 \quad .$$

In addition to (12) appropriate estimates on e and s are postulated. In particular they imply that the main terms of e are positive and contain ψ_0 and ψ_1, and that s does not contain the main terms of e (see [3]).

The standard example of a free energy φ that satisfies the above structural assumptions is

$$(13) \qquad \varphi(u, w) := \alpha_V \varphi_*(w) + \alpha_1(1 - w\theta_{crit})u^2 + \alpha_2 w|u|^\nu$$

with constants $\theta_{crit} > 0$, $\alpha_V, \alpha_1, \alpha_2 > 0$, and $\nu > 2$. The φ_*-term is the strictly concave term of φ. It refers to pure heat conduction and in particular can be postulated as

$$\varphi_*(w) = \log w, \quad \text{or} \; = -\frac{1}{w}, \quad \text{or} \; = -\log(1 + \frac{1}{w}) \quad .$$

where the first form refers to a constant heat coefficient, and for the second and third form the third law of thermodynamics $s(u, w) \to 0$ as $w \to \infty$ is satisfied.

We now present the existence result for system (1)-(3) with initial and boundary conditions (9), (10). The proof is given in [3] and is based on an implicit time-discrete approximation of the problem. The main tool used in [3] is the Clausius-Duhem inequality for the entropy production which plays the role of a parabolic energy estimate. It is applied to the time discrete solution. Together with parabolic estimates which show that the time-discrete solutions are compact in $L^1(\Omega_{t_0})$ (using ideas in [1]) it follows that they converge to a weak solution of the system.

The notion of a weak solution to the system (1)-(3) with conditions (9), (10) is given by three differential operators $\mathcal{F}_k, k = 0, 1, 2$, defined as follows:

$$\langle \zeta, \mathcal{F}_0(u, v, w) \rangle := \int_{\Omega_{t_0}} \zeta(-v + \varphi_{,u}(u, w)) + \int_{\Omega_{t_0}} \kappa w \nabla \zeta \cdot \nabla u \quad ,$$

$$\langle \xi, \mathcal{F}_1(u, v, w) \rangle := -\int_{\Omega_{t_0}} \partial_t \xi(u - u_0) - \int_{\Omega_{t_0}} \nabla \xi \cdot \vec{j}(u, v, w) \quad ,$$

$$\langle \eta, \mathcal{F}_2(u, v, w) \rangle := \int_{\Omega_{t_0}} \partial_t \eta(E(u, w) - E(u_0, w_0))$$

$$+ \int_{\Omega_{t_0}} (\nabla \eta \cdot \vec{q}(u, v, w) + \eta g) + \int_{\Gamma_{t_0}} \eta p(u, v, w) d\mathcal{H}^n$$

for test functions $(\zeta, \xi, \eta) \in L^2(]0, t_0[; V)$, $V := H^{1,2}(\Omega; \mathbb{R}^3)$, satisfying

$$\zeta \in L^\infty(\Omega_{t_0}) \quad \text{and} \quad \sqrt{w}\nabla\zeta \in L^2(\Omega_{t_0}) \quad ,$$

$$\partial_t\xi \in L^2(\Omega_{t_0}) \quad \text{with} \quad \xi(t_0) = 0 \quad ,$$

$$\eta, \partial_t\eta \in L^\infty(\Omega_{t_0}) \quad \text{with} \quad \eta(t_0) = 0 \quad .$$

These operators are defined for triples $(u, v, w) \in L^2(]0, t_0[; V)$ for which in addition $\varphi_{,u}(u, w), w|\nabla u|^2$, and $E(u, w)$ are in $L^1(\Omega_{t_0})$. In [3] the following existence result is shown.

Theorem 1. There exists $(u, v, w) \in L^2(]0, t_0[; V)$ with the properties

$$\varphi_{,u}^0(u, w), \varphi_{,u}^1(u, w), u\varphi_{,u}^0(u, w), w|\nabla u|^2 \in L^1(\Omega_{t_0}) \quad ,$$

$$\varphi_{,w}(u, w), |\nabla u|^2 \in L^\infty(]0, t_0[; L^1(\Omega)) \quad ,$$

and there exisits a non-negative bounded measure λ on $\bar{\Omega}_{t_0}$, so that for all test functions (ξ, ξ, η) as above with $\partial_t\eta \in C^0(\bar{\Omega}_{t_0})$

$$0 = \langle \zeta, \mathcal{F}_0(u, v, w) \rangle, \quad 0 = \langle \xi, \mathcal{F}_1(u, v, w) \rangle \quad ,$$

$$0 = \langle \eta, \mathcal{F}_2(u, v, w) \rangle + \int_{\bar{\Omega}_{t_0}} \partial_t\eta d\lambda \quad .$$

We call (u, v, w) a weak solution to the system (1)-(3), (9), (10) if these equations are satisfied with $\lambda = 0$. It can be seen, that the support of λ is concentrated on $\{w = 0\}$. Since $w > 0$ almost everywhere with respect to the $n + 1$- dimensional Lebesgue measure on Ω_{t_0}, this means that λ has no Lebesgue density. Of course, if the system would provide a maximum principle for w from below, this would imply that $\lambda = 0$. However, the surface energy term $\frac{\kappa}{2}|\nabla u|^2$ in the internal energy $E(u, w)$, which is independent of w, is a strong indication against such a maximum principle. Therefore a different procedure is necessary to show that $\lambda = 0$. In [3] the following positive answer has been found using additional growth conditions on the potential φ.

Theorem 2. Assume that $l_{22} = 1, l_{21} = 0$, and that $n \leq 3$. Then the triple (u, v, w) from Theorem 1 satisfies

$$\varphi_{,w}(u, w), |\nabla u|^2 \in L^\infty(]0, t_0[, L^2(\Omega))$$

and it follows that $\lambda = 0$. Therefore (u, v, w) is a weak solution.

References:

[1] H.W. Alt, S. Luckhaus, Quasilinear elliptic-parabolic differential equations, Math. Z. **183** (1983), 311-341.

[2] H.W. Alt, I. Pawlow, A mathematical model of dynamics of non-isothermal phase separation, SFB 256 Univ. Bonn, Preprint No 158, 1991.

[3] H.W. Alt, I. Pawlow, Existence of solutions for non-isothermal phase separation, SFB 256 Univ. Bonn, Preprint No 159, 1991.

[4] J.W. Cahn, On spinodal decomposition, Acta Metall. **9** (1961), 795-801.

[5] J.W. Cahn, J.E. Hilliard, Free energy of a nonuniform system. I. Interfacial free energy, J. Chem. Physics **28** (1958), 258-267.

[6] C.M. Elliott, The Cahn-Hilliard model for the kinetics of phase separation, in: "Mathematical Models for Phase Change Problems", J.F. Rodrigues (Ed.) International Series of Numerical Mathematics, Vol. 88. Birkhäuser-Verlag, Basel, 1989, 35-73.

[7] C.M. Elliott, D.A. French, Numerical studies of the Cahn-Hilliard equation for phase separation, I.M.A. Journal Appl. Math. **38** (1987), 97-128.

[8] C.M. Elliott, Zheng Songmu, On the Cahn-Hilliard equation, Arch. Rat. Mech. Anal. **96** (1986), 339-357.

[9] T. Miyazaki, T. Kozakai, S. Mizuno, M. Doi, A theoretical analysis of the phase decompositions based upon the non-linear diffusion equation, Trans. Japan Inst. Metals **24** (1983) , 246-254.

[10] A. Novick-Cohen, L.A. Segel, Nonlinear aspects of the Cahn-Hilliard equations, Physica D **10** (1984), 277-298.

[11] W. v. Wahl, On the Cahn-Hilliard equation $u' + \Delta^2 u - \Delta f(u) = 0$, Delft Progress Report (1985) **10**, 291-310.

[12] Zheng Songmu, Asymptotic behavior of the solution to the Cahn-Hilliard equation, Applic. Anal. **23** (1986), 165-184.

Hans Wilhelm Alt
Institut für Angewandte Mathematik
Wegelerstr. 6, D-5300 Bonn, W. Germany

Irena Pawlow
Systems Research Institute, Polish Academy of Sciences
Newelska 6, PL-01-447 Warsaw, Poland

P W BATES

Coarsening and nucleation in the Cahn–Hilliard equation

This represents joint work with N. Alikakos and G. Fusco [ABF] and work in progress with P. Fife [BF2]. We consider a binary alloy at a fixed temperature occupying a region Ω. If two constituents are present in amounts u_1 and u_2, the concentration of the first is given by $u = u_1/(u_1 + u_2)$. There are two specific concentrations, $\alpha < \beta$, such that alloy with concentration α can coexist in contact with alloy at concentration β, and if the average concentration lies in (α, β), then the material tends to form into a granular structure with regions having concentration α separated by regions having concentration β. In addition, surface tension or neighbor effects tend to preclude arbitrarily fine grained and favor coarse grained structure.

A reasonable model for the free energy is given by

$$(1) \qquad J(u) = \int_{\Omega} \left\{ \frac{\varepsilon^2}{2} \left| \nabla u \right|^2 + W(u) \right\} dx,$$

where W is a double well potential with minima at α and β and ε is a small parameter related to molecular interaction length.

The dynamical behavior of the concentration field u should be governed by the steepest descent of the free energy functional:

$$(2) \qquad u_t = -\text{grad } J(u)$$

For this equation to make sense (recall that the gradient of a functional on a space V is a member of the dual space V') we must interpret the gradient as a *representative* of that functional which lies in V. Thus, we are forced to restrict our state space to being Hilbert space. Still, there is some freedom here in how to interpret (2). The natural choice may be $V = L^2(\Omega)$ in which case, after calculating the representative of grad $J(u)$, we see that (2) may be written as

(3)
$$\begin{cases} u_t = \varepsilon^2 \Delta u - W'(u) & \text{in } \Omega \\[2mm] \dfrac{\partial u}{\partial v} = 0 & \text{on } \partial\Omega. \end{cases}$$

While this is an interesting equation which models some aspects of the phenomenon being modelled (see e.g. [B], [BK], [dMS], [CP], [FH]) it fails to conserve average concentration $\dfrac{1}{|\Omega|} \displaystyle\int_{\Omega} u \, dx$. The simplest model which does not have that shortcoming arises by taking $V = H^{-1}$, the dual of the Sobolev space H^1. Calculating the representative of grad $J(u)$ in H^{-1} shows that (2) may be written as

(4)
$$\begin{cases} u_t = \Delta(-\varepsilon^2 \Delta u + W'(u)) & \text{in } \Omega \\[2mm] \dfrac{\partial u}{\partial v} = \dfrac{\partial \Delta u}{\partial v} = 0 & \text{on } \partial\Omega. \end{cases}$$

This is the Cahn-Hilliard equation (see [C], [CH]). This derivation brings more than just a sense of harmony to the well-known fact that (1) is a Lyapunov functional for the Cahn-Hilliard equation, it tells us that the gradient structure of the equation is properly exploited by working in H^{-1}.

This has been done to some extent in order to analyze the spectrum of the linearized equation [BF1]. In [ABF] the one dimensional equation is considered and the above observation regarding H^{-1} provides the impetus to study the equivalent integrated equation:

(5)
$$\begin{cases} \tilde{u}_t = -\varepsilon^2 \tilde{u}_{xxxx} + (W'(\tilde{u}_x))_x, & 0 < x < 1 \\ \tilde{u}(0,t) = 0, \ \tilde{u}(1,t) = M, \ \tilde{u}_{xx}(0,t) = 0 = \tilde{u}_{xx}(1,t), \end{cases}$$

where $\tilde{u}(x) \equiv \int_0^x u(y)dy$ and M is the average concentration of the initial state u_0.

Consider the case where M lies in (α, β). The main result in [ABF] concerns the stationary solution u^* which has two interior transition layers (see Figure 1).

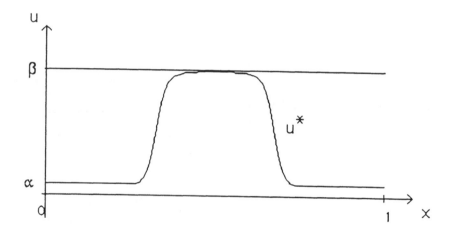

Figure 1

By working mainly with (5) and its linearization we prove

Theorem 1 Viewing (4) as dynamical system

(a) u^* has a one dimensional unstable manifold $\mathfrak{M}$ which is Lipschitz in $H^{4-\delta}$.

(b) This manifold is globally approximated in H^2 by translates of (the periodically extended) u^*, the accuracy of approximation being of order $e^{-d/\varepsilon}$ for some fixed $d > 0$.

(c) The transition layers of functions in $\mathfrak{M}$ move with speed of order $e^{-d/\varepsilon}$.

This extremely slow motion has been observed in the reaction-diffusion equation (3) (see e.g. [CP], [FH]) and has been called metastability.

If u* were to have n layers, then we expect it to have an $(n-1)$-dimensional unstable manifold with the speed on that manifold, as measured by the motion of the transition layers, being of order $e^{-d/n\varepsilon}$. Thus, starting with a small perturbation of a multilayered stationary state, there would be a succession of extremely slow but progressively slower time scales during which the phase structure coarsens.

NUCLEATION

The second topic discussed concerns the case when $\alpha < M < \beta$ with $W''(M) > 0$. For such a concentration the homogeneous state is stable but is not a global minimum for the free energy (1). Perturbations of sufficient intensity, while preserving the average concentration, can destabilize the homogeneous state leading towards a two phase energy minimizing solution.

Here we consider spatially localized perturbations, examine the threshold between decay and propagation of the disturbance, and analyze the nature of that motion. The first observation is that besides the homogeneous stable solution, there are two other monotone increasing solutions with this average concentration (see Figure 2).

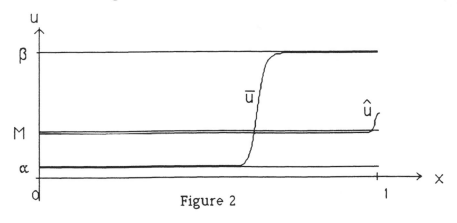

Figure 2

It is conjectured that the solution with the boundary layer, $\hat{u}$, has a one dimensional unstable manifold leading to either the homogeneous state or the energy minimizing single-layered state, $\bar{u}$. The stable manifold of $\hat{u}$ separates the basins of attraction of these two stable solutions. We call $\hat{u}$ the nucleation stationary state. Note that interior "blip" stationary states can be found by reflection and rescaling space.

Even though $\hat{u}$ and $\bar{u}$ are monotone, points on the one dimensional orbit connecting these states need not be spatially monotone functions. Considering how a nucleation front would propagate and formal asymptotics suggest that monotonicity is destroyed by a single diffusive front developing into the interior from the boundary layer (see Figure 3). This conjecture was subsequently reinforced by numerical evidence.

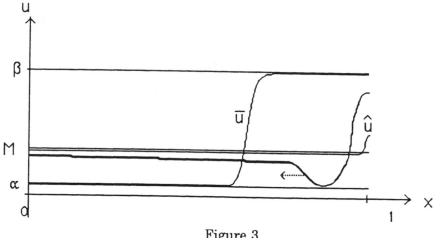

Figure 3

REFERENCES

[ABF] N. D. Alikakos, P. W. Bates and G. Fusco, *Slow motion for the Cahn-Hilliard equation in one space dimension*, <u>J. Diff. Eqs.</u>, to appear.

[BF1] P. W. Bates and P. C. Fife, *Spectral comparison principles for the Cahn-Hilliard and Phase-Field equations and time scales for coarsening*, <u>Physica D.</u>, to appear.

[BF2] P. W. Bates and P. C. Fife, *Nucleation in the Cahn-Hilliard equation*, in preparation.

[B] L. Bronsard, Ph.D. dissertation, Courant Institute, 1988.

[BK] L. Bronsard and R. V. Kohn, *On the slowness of phase boundary motion in one space dimension*, <u>Comm. Pure Appl. Math.</u>, to appear.

[C] J. W. Cahn, *On spinodal decomposition*, <u>Acta Met</u> 9, (1961) 795-801.

[CH] J. W. Cahn and J. E. Hilliard, *Free energy of a nonuniform system I. Interfacial free energy*, <u>J. Chem. Phys.</u> 28, (1958) 258-267.

[CGS] J. Carr, M. Gurtin and M. Slemrod, *Structured phase transitions on a finite interval*, <u>Arch. Rat. Mech. Anal.</u> 86, (1984) 317-351.

[CP] J. Carr and R. L. Pego, *Metastable patterns in solutions of* $u_t = \varepsilon^2 u_{xx} - f(u)$, <u>Comm. Pure Appl. Math.</u>, 42, (1989) 523-576.

[dMS] P. de Mottoni and M. Schatzman, *Evolution geometrique d'interfaces*, <u>C. R. Acad. Paris</u> (1988).

[FH] F. Fusco and J. K. Hale, *Slow motion manifolds, dormant instability and singular perturbations*, <u>J. Dynamics and Diff. Eqs.</u> 1 (1989) 75-94.

Peter W. Bates
Brigham Young University

H HATTORI AND K MISCHAIKOW

On a global dynamics of a phase transition problem*

Abstract

In this note we discuss a dynamical systems approach to a phase transition based on the Korteweg theory of capillarity. First we discuss the existence of solutions. Then, we discuss the bifurcation diagram of stationary solutions and their stability. Lastly, we study the connecting orbit problems in the semiflow.

1 Introduction

In this note we study the global dynamics for the following system of parabolic equations

$$p_t = \nu p_{xx} - \eta q_{xx} + \sigma(q) - P,$$

$$q_t = p_{xx},$$

$$(1.1)$$

*The first author was supported by Army Grant DAAL 03-89-G-0088 and the second author was supported by a NSF grant.

where $x \in [0, 1]$ and the boundary conditions are given by

$$p_x(0, t) = 0, \quad p(1, t) = 0,$$
$$q_x(0, t) = 0, \quad q_x(1, t) = 0. \tag{1.2}$$

The above system is derived from an equation

$$u_{tt} = \sigma(u_x)_x + \nu u_{xxt} - \eta u_{xxxx} \tag{1.3}$$

with boundary conditions

$$u(0, t) = 0, \quad \sigma(u_x(1, t)) + \nu u_{xt}(1, t) - \eta u_{xxx}(1, t) = P, \tag{1.4}$$

$$u_{xx}(0, t) = 0, \quad u_{xx}(1, t) = 0, \tag{1.5}$$

by setting $p = \int_1^x u_t \, dx$ and $q = u_x$. Equation (1.3) models a bar which goes through the phase transition. The boundary conditions (1.4) shows that the bar is under the soft loading device. The boundary conditions (1.5) are the natural boundary conditions for the corresponding variational problem. The terms with the coefficients ν and η are called viscosity and capillarity terms, respectively. In what follows, we assume that σ is given by Fig. 1.1. In this figure $(0, \alpha^*]$ and $[\beta^*, \infty)$ are called the α-phase and the β-phase, respectively. They correspond to the different phases of materials. Andrew & Ball [AB] and Slemrod [S] considered equation (1.3). Pego [P], Andrew & Ball [A,AB], and Defermos [D] considered the equation

$$u_{tt} = \sigma(u_x)_x + \nu u_{xxt}. \tag{1.6}$$

Pego has shown that some discontinuous stationary solutions are dynamically stable. In what follows, we discuss first the existence of global solutions, then stability of stationary solutions and the bifurcation diagram, and finally the connecting orbit problems in the semiflow. We omit the proofs, as they will appear in [HM].

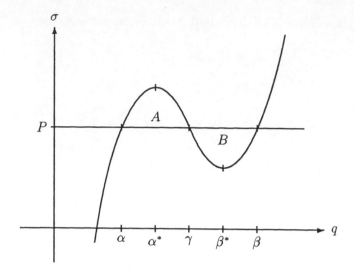

Figure 1.1

2 Existence

We state the theorem establishing the existence of a global solution to (1.1) and (1.2).

Theorem 2.1 *Suppose σ is Lipshitz continuous and that $(p_o, q_o) \in H^1(0,1)$. Then, there exists a unique global solution $(p,q) \in H^2(0,1)$.*

We define the operator A by

$$A \begin{pmatrix} p \\ q \end{pmatrix} = \begin{pmatrix} \nu p_{xx} - \eta q_{xx} \\ p_{xx} \end{pmatrix}$$

and show that A with the boundary conditions (1.2) is an infinitesimal generator of a compact analytic semigroup. It should be mentioned that since the boundary conditions are not typical, it causes some difficulty in estimating the eigenvalues and the resolvent. The compact analytic semigroup gives a local existence. Now, we use the following identity

$$E(p, q)(t) + \int_0^t \int_0^1 \nu p_{xx}^2(x, s) \, dx \, ds = E(p, q)(0),$$ (2.7)

where

$$E(p, q)(t) = \int_0^1 \{\frac{1}{2}p_x^2 + W(q) - Pq + \frac{\eta}{2}q_x^2\}(x, t) \, dx,$$ (2.8)

as the *a priori* estimate for the H^1 norm of (p, q) to show the existence of a glabal solution.

3 Dynamics

Here, we discuss the stability of stationary solutions and the bifurcation diagram.

Lemma 3.1 *The constant solutions* $(0, \alpha)$, $(0, \beta)$, *and* $(0, \delta)$ *are stationary solutions for all values of* $\eta > 0$. *Furthermore, their indices are* $h((0, \alpha)) = h((0, \beta)) = \Sigma^0$. *Namely, they are dynamically stable.*

Next, consider the eigenvalue problem corresponding (1.1) and (1.2):

$$\eta q_{xx} - (\nu\lambda + \sigma'(\delta))q = -\lambda,$$

$$p_{xx} = \lambda q.$$ (3.9)

Lemma 3.2 *The eigenvalues of (3.9) cross the origin from left to right of the imaginary axis at* $\eta = -\sigma'(\delta)/(n\pi)^2$ *as we decrease* η. *Furthermore zero eigenvalues are simple.*

Lemma 3.3 *If* $\sigma \in C^3$, $\sigma'(\delta) < 0$, *and* $\sigma'''(\delta) > 0$, *then there is a supercritical pitchfork bifurcation at* $\eta = -\sigma'(\delta)/(n\pi)^2$. *Furthermore, if* $\sigma(u + \delta)/u < \sigma'(\delta)$ *for* $\alpha - \delta \leq u \leq \beta - \delta$ *except at* $u = 0$, *then there is no secondary bifurcation along the non-constant stationary solutions.*

Theorem 3.1 *There exists a global compact attractor for (1.1) and (1.2).*

The above lemmas and the theorem imply

Lemma 3.4 *If $\eta > -\sigma'(\delta)/\pi^2$, the stationary solution $(0,\delta)$ has one dimensional unstable manifold or equivalentely the index is $\mathbf{h}((0,\delta)) = \Sigma^1$.*

Combining the above lemmas and the theorem we have

Theorem 3.2 *If $\sigma \in C^3$, $\sigma'(\delta) < 0$, and $\sigma'''(\delta) > 0$, then the following holds:*

(i) For $-\sigma'(\delta)/(n\pi)^2 < \eta < -\sigma'(\delta)/((n-1)\pi)^2$, $(0,\delta)$ is a nondegenerate stationary solution and has an n-dimensional unstable manifold.

(ii) If $M(k^\pm)$ denote the non-constant stationary solutions which arise from the bifurcation point $\eta = -\sigma'(\delta)/(k\pi)^2$, then $M(k^\pm)$ are non-degenarate and have k-dimensional unstable manifolds.

4   Connecting orbit problems

We now discuss the connecting orbit problem in the semiflow. In the semiflow the connecting orbit means the solutions connecting two stationary solutions, namely,

$$\lim_{t\to-\infty}(p,q)(t) = \text{a stationary solution},$$

$$\lim_{t\to\infty}(p,q)(t) = \text{another stationary solution}.$$

To simplify the notation let $M(0^+) = (0,\alpha)$, $M(0^-) = (0,\beta)$, and $M(n) = (0,\delta)$. Then, we have

Theorem 4.1 *Given a collection $\{j^*, j+1^*, j+2^*, ..., j+r^* \mid * = +or-\}$ and $\epsilon > 0$, there exists a solution $(p(t), q(t))$ of (1.1) and (1.2) and a sequence $t_1 > t_2 > ... > t_{r-1}$ such that*

$$\lim_{t\to-\infty}(p(t),q(t)) = M(j+r^*), \quad \lim_{t\to\infty}(p(t),q(t)) = M(j^*), \qquad (4.10)$$

and

$$d(M(j + i^*), (p(t_i), q(t_i))) < \epsilon. \tag{4.11}$$

Furthermore,

$$cl(C(M(j + r^*), M(j^*))) \cap cl(C(M(j + i^*), M(j + s^*))) \neq \emptyset, \ for \ 0 \leq s \leq r.$$

This theorem establishes that there is always a connecting orbit from a stationary solution with higher dimensional unstable manifold to lower one. To prove this we use the connection matrix, which is an extension of the Conley's index.

References

[A] Andrews, G., On the existence of solutions to the equation $u_{tt} = u_{xxt} + \sigma(u_x)_x$, J. Deff. Eqns. 35 (1980), 200-231.

[AB] Andrews, G. and J.M. Ball, Asymptotic behaviour and change of phase in one-dimensional nonlinear viscoelasticity, J. Diff. Eqns. 44 (1982), 306-341.

[D] Dafermos, C.M., The mixed initial-boundary value problem for the equation of nonlinear one-dimensional viscoelasticity, J. Diff. Eqns. 6 (1969), 71-86.

[HM] Hattori, H. and C. Mischaikow, A dynamical systems approach to a phase transition problem, to appear in J. Diff. Eqns.

[P] Pego, R., Phase transitions: Stability and admissibility in one-dimensional viscoelasticity, IMA Report # 180 (1985).

[S] Slemrod, M., Admissiblity criteria for propagating phase boundaries in a van der Waals fluid, Arch. Rat. Mech. Anal. 81 (1983), 301-315.

Harumi Hattori
Department of Mathematics
West Virginia University
Morgantown, WV 26506

Konstantin Mischaikow
School of Mathematics
Geogia Institute of Technology
Atlanta, GA 30332

D HILHORST[1], F ISSARD-ROCH[1] AND T I SEIDMAN[2]

On a reaction-diffusion equation with a free boundary: the case of an unbounded domain

1. Introduction

In heterogeneous media, most chemical reactions result in the creation or consumption of one or more phases. For example, in liquid-solid systems, the solid phase consists of small grains of a single compound and the liquid phase is a dilute solution of that compound. Such systems appear for instance in chemistry or in geology.

We consider here a simple model of growth and dissolution in an unbounded domain [1] : a single spherical grain of radius $R(t)$ is surrounded by a liquid phase. We denote by $C(r,t)$, $r > R(t)$, the concentration in the liquid phase, where r is the distance from the center of the grain. Then C and R satisfy the following rescaled equations

$$(P) \begin{cases} C_t = \dfrac{1}{r^2}(r^2 C_r)_r + F(C), \ t > 0, \ r > R(t) & (1.1) \\[2mm] C_r = (1 - C)H(r, C), \ t > 0, \ r = R(t) & (1.2) \\[2mm] \dot{R}(t) = H(R(t), C(R(t), t)), \ t > 0 & (1.3) \\[2mm] R(0) = R_0 > 0 & (1.4) \\[2mm] C(r, 0) = C_0(r), \ r > R_0 & (1.5) \end{cases}$$

where
. $C_0 - \Gamma \in H^1(R_0, +\infty)$ for some constant $\Gamma \in [0, 1]$, $0 \le C_0 \le 1$ in $(R_0, +\infty)$,
. $F \in C^1(\mathbb{R})$ is nonincreasing, and $F(\Gamma) = 0$,
. $H(r, w) = g_1(w) - g_2(r)$, $g_1 \in C^{0,1}(\mathbb{R})$, $g_1(0) \le 0$; $g_2 \in C^{0,1}(\mathbb{R}^+)$, $g_2 \ge 0$,
$\lim\limits_{r \to +\infty} g_2(r) = g_*$ where g_* is a positive constant.

([1]) Laboratoire d'Analyse Numérique, Université Paris-Sud, Bâtiment 425, 91405 Orsay, France.
([2]) Department of Mathematics, University of Maryland Baltimore County, Baltimore, MD 21228, U.S.A.

Equation (1.1) describes the mass balance in the liquid phase, whereas equation (1.2) corresponds to the mass balance at the solid-liquid interface. The dissolution and growth of the grain is expressed in equation (1.3). In the physical case, the process follows Nernst's law, namely $H(r, w) = K(w/V - G(r))$ with $G(r) = C_* \ exp(\frac{\delta}{r})$. Here $1/V$ denotes the concentration in the solid phase, C_* the concentration at saturation of the solution and δ is proportional to the surface tension of the interface. However one expects that this expression of $G(r)$ is no longer valid for small r.

The purpose of this note is to prove that problem (P) is well posed. More precisely, we prove that Problem (P) has a unique solution (R, C) on a maximum time interval $(0, \tau_*)$ which is such that either $\tau_* = +\infty$ or $R(t) \to 0$ as $t \to \tau_*$. We also give a lower bound for τ_* and some conditions which guaranty that $R(t)$ remains bounded.

This work extends results of F. Conrad, D. Hilhorst and T.I. Seidman on a bounded domain [2,5]. For more details, we refer to [4].

2. Preliminaries.

Let $R_0 > 0$ be fixed. We first modify functions F and H outside of the domain $\{(\omega, r) \in [0, 1] \times [R_0/2, +\infty)\}$ in such a way that F and H are bounded and denote again by F and H the modified functions. Then there exists $\tau_0 > 0$ such that if (R, C) is a solution of (P) on $[0, \tau_0]$ with $0 \leq C \leq 1$, then $R(t) \geq \frac{R_0}{2}$ for all $t \in (0, \tau_0)$.

One first shows that Problem (P) has a unique solution (R, C) on $(0, \tau_0)$ and then characterizes the maximum time of existence τ_*. The method of proof is based on the use of a fixed point theorem in a suitable weighted space. More precisely we introduce the map T defined as follows:

i) Let $\omega \in L^2(0, \tau_0)$; then the initial value problem :

$$(IVP) \quad \begin{cases} \dot{R}(t) = H(R(t), \omega(t)), & t \in (0, \tau_0) \\ R(0) = R_0 \end{cases}$$

has a unique solution $R \in C^{0,1}([0, \tau_0])$ and we have $R(t) \geq \frac{R_0}{2}$ for all $t \in [0, \tau_0]$.
ii) The next step is to prove that the Problem (1.1),(1.2),(1.5) where R is the function obtained in (i) has a unique solution C; we denote this problem by $(\hat{P})$. However, since it will be necessary further in the proof to compare two solutions C_1 and C_2 on the two corresponding boundaries R_1 and R_2, it is useful to transform problem $(\hat{P})$ into a problem on a fixed domain. Set $y = r - R(t)$,

$u(y,t) = C(r,t)$; Problem $(\hat{P})$ becomes

$$(\hat{P}_1) \quad \begin{cases} u_t = u_{yy} + u_y \psi(y,t) + F(u), \ y \geq 0, \ t \in (0,\tau_0) & (2.1) \\ u_y(0,t) = \Phi(t,u(0,t)), \ t \in (0,\tau_0) & (2.2) \\ u(y,0) = u_0(y), \ y \geq 0 \end{cases}$$

with

· $\psi(y,t) = \dot{R}(t) + \frac{2}{y+R(t)}$, $y \in (0,\infty)$, $t \in (0,\tau_0)$
· $\Phi(t,z) = (1-z)\hat{H}(R(t),z)$, $z \in \mathbb{R}$, $t \in (0,\tau_0)$
· $u_0(y) = C_0(y+R_0)$, $y \in \mathbb{R}^+$.

Note that the free boundary $r = R(t)$ has been transformed into the line $y = 0$.

iii) Finally, one sets :

$T(\omega)(t) = u(0,t)$ for all $t \in (0,\tau_0)$.

Now suppose that ω is a fixed point of T; let R satisfy the initial value problem (IVP), u satisfy Problem $(\hat{P}_1)$ and $C(r,t) = u(y,t)$ with $r = y + R(t)$, then (R,C) is a solution of Problem (P) on $(0,\tau_0)$.

3. Existence and uniqueness of the solution of (P) on $(0,\tau_0)$.

A first step in the proof is to show that Problem $(\hat{P}_1)$ is well-posed. This is expressed in the following result.

Theorem 1. Let $R \in C^{0,1}([0,\tau_0])$ be such that $R(t) \geq \frac{R_0}{2}$ for all $t \in [0,\tau_0]$ and $R(0) = R_0$; then, if $u_0 - \Gamma \in H^1(\mathbb{R}^+)$, there exists a unique (weak) solution u of Problem $(\hat{P}_1)$, which is such that $u - \Gamma \in L^\infty(0,\tau_0; H^1(\mathbb{R}^+)) \cap L^2(0,\tau_0; H^2(\mathbb{R}^+))$ and $u_t \in L^2(\mathbb{R}^+ \times (0,\tau_0))$.

In order to show that T is a strict contraction in a suitable weighted space, one gives estimates for differences of solutions of Problem $(\hat{P}_1)$.

Theorem 2. Let $\omega_1, \omega_2 \in L^2(0,\tau_0)$, R_1 and R_2 be the corresponding solutions of the initial value problem (IVP) and let u_1 and u_2 be the solutions of Problem $(\hat{P}_1)$ corresponding to R_1 and R_2. There exist constants D and $\lambda_0 > 0$ such that

$$e^{-\lambda t} \parallel (u_1 - u_2)(\cdot,t) \parallel^2 + \frac{\lambda}{2} \int_0^t \parallel (u_1 - u_2)(\cdot,s) \parallel^2 e^{-\lambda s} ds$$

$$+ \int_0^t \parallel (u_1 - u_2)_y(\cdot,s) \parallel^2 e^{-\lambda s} ds$$

$$\leq D \int_0^t e^{-\lambda s} (w_1(s) - w_2(s))^2 ds$$

for all $\lambda > \lambda_0$ and $t \in [0, \tau_0]$, where $\| \cdot \|$ denotes the norm in $L^2(\mathbb{R}^+)$.

Next we define the following norm on $L^2(0, \tau_0)$:

$$\| w \|_\lambda = \int_0^{\tau_0} e^{-\lambda s} \omega^2(s) ds, \quad for \ \omega \in L^2(0, \tau_0);$$

and state a result which will be useful in the sequel.

Lemma 3. For each $\epsilon > 0$, there exists a positive constant C_ϵ such that for all $v \in H^1(\mathbb{R}^+)$,

$$\|v\|^2_{C([0,\infty])} \leq \epsilon \|v'\|^2_{L^2(\mathbb{R}^+)} + C_\epsilon \|v\|^2_{L^2(\mathbb{R}^+)}.$$

We are now ready to prove the following contraction result.

Theorem 4. For λ sufficiently large, T defines a strict contraction on $L^2(0, \tau_0)$ with the norm $\| \cdot \|_\lambda$.

Proof. It follows from Lemma 3 that, for each $\epsilon > 0$, there exists $C_\epsilon > 0$ such that

$$\|T\omega_1 - T\omega_2\|^2_\lambda \leq \epsilon \int_0^{\tau_0} e^{-\lambda s} \|(u_1 - u_2)_y(\cdot, s)\|^2 ds$$

$$+ C_\epsilon \int_0^{\tau_0} e^{-\lambda s} \|(u_1 - u_2)(\cdot, s)\|^2 ds,$$

from which we deduce, using also Theorem 3 that

$$\|T_{\omega_1} \leq T_{\omega_2}\|^2_\lambda \leq D \left(\epsilon + \frac{2C_\epsilon}{\lambda} \right) \|\omega_1 - \omega_2\|^2_\lambda$$

$$\leq \frac{1}{2} \|\omega_1 - \omega_2\|^2_\lambda$$

for $\epsilon \leq \frac{1}{4D}$ and $\lambda \geq 8C_\epsilon D$.

It then easily follows that Problem (P) has a unique solution (R, C) on $[0, \tau_0]$ such that $0 \leq C \leq 1$. Furthermore $R \in C^1([0, \tau_0])$.

4. The maximum time of existence.

Let $\mathcal{A}$ be the set of the triples (R, C, τ) such that $\tau > 0$ and

i) $R \in C^{0,1}([0, \tau])$ and $R(t) > 0$ for all $t \in [0, \tau]$,

(ii) (R, C) is a solution of Problem (P) on $[0, \tau]$ such that $0 \leq C \leq 1$.

Then $\mathcal{A}$ is ordered by τ and we define

$$\tau_* = sup\{\tau > 0 \ such \ that \ there \ exists \ (R, C, \tau) \in \mathcal{A}\}.$$

235

The previous local analysis implies that $\tau_* > 0$; possibly $\tau_* = +\infty$ depending on the data.

__Theorem 5.__ Either $\tau_* = +\infty$, or $\tau_* < +\infty$ and $R(t) \to 0$ as $t \to \tau_*$.
. If g_1 is nondecreasing then

$$\tau_* \geq \int_0^{R_0} \frac{dr}{g_2(r) - g_1(0)}.$$

Next we give some assumptions on the functions g_1 and g_2 which guaranty that if R_0 is not too large, $R(t)$ remains bounded.
H_g : g_1 is non decreasing and $g_1(1) > g_*$,
$\quad g_1(1) < \|g_2\|_{L^\infty(\mathbb{R}^+)} < +\infty$
$\quad$ Let $r_{**} = sup\{r > 0, \ g_2(r) = g_1(1)\}$; there exists $r_* \in (0, r_{**})$ such that $g_2(r_*) = g_1(1)$ and $g_2(r) > g_1(1)$ for all $r \in (r_*, r_{**})$.

__Theorem 6.__
(i) Let $R_0 \leq r_*$ then $R(t) \leq r_*$ for all $t \in (0, \tau_*)$;
(ii) If $R_0 \in (r_*, r_{**})$ then there exists $\tau_1 \in (0, \tau_*)$ such that $\dot{R}(t) < 0$, $r_* < R(t) < R_0$ for all $t \in (0, \tau_1)$ and $R(t) \leq r_*$ for all $t \in (\tau_1, \tau_*)$.

References.

[1] F. Conrad and M. Cournil, Free boundary problems in dissolution growth processes, in Boundary Control and Boundary Variations, J.P. Zolesio Ed., Lecture Notes in Control and Information Sciences, 100 (1988) 116-136.
[2] F. Conrad, D. Hilhorst and T.I. Seidman, Well-posedness of a moving boundary problem arising in a dissolution-growth process, to appear in Nonlinear Analysis T.M.A.
[3] J. Hale, Ordinary Differential Equations, R.E. Krieger Publishing Company, New-York (1980).
[4] D. Hilhorst, F. Issard-Roch and T.I. Seidman, On a free boundary problem arising in a dissolution-growth process : The case of an unbounded domain, in preparation.
[5] T.I. Seidman, Optimal control and well-posedness for a free boundary problem, these Proceedings.

236

J SPREKELS[1]

On thermomechanical phase transitions

1 Introduction

In this paper we consider thermomechanical processes in one–dimensional heat–conducting solids of constant density ρ under heating and loading. We think of metallic solids that do not only respond to a change of the strain $\epsilon = u_x$ (u stands for the displacement) by an elastic stress $\sigma = \sigma(\epsilon)$, but also react to changes of the curvature of their metallic lattices by a couple stress $\mu = \mu(\epsilon_x)$. Thus, the corresponding free energy density F is assumed in the Ginzburg–Landau form $F = F(\epsilon, \epsilon_x, \theta)$, where θ is the absolute temperature. In the framework of the Landau theory of phase transitions, ϵ plays the role of the *order parameter* whose actual value determines what phase is prevailing in the material (see [3]). Since we are interested in solid–solid phase transitions, driven by loading and/or heating and accompanied by hysteresis effects, we do not assume that $F(\cdot, \epsilon_x, \theta)$ is convex.

Particularly interesting materials are metallic alloys like *CuZn*, *CuSn*, *AuCuZn*, *AgCd*, *TiNi* which exhibit the so–called *shape memory effect*. In these materials the metallic lattice is deformed by shear, and the assumption of a constant density is justified. The relations between shear stress and shear strain ($\sigma - \epsilon$–curves) show a temperature–dependent hysteresis (see Fig.1). For an account of the properties of shape memory materials we refer to [2].
On the microscopic scale this behaviour is ascribed to first–order stress– or temperature–induced phase transitions between different configurations of the crystal lattice, namely the symmetric high–temperature phase *austenite* (taken as reference configuration) and its sheared counterpart *martensite* which prevails in two oppositely oriented versions at low temperatures (cf., [6,7]).

The simplest form for F which matches the experimental evidence quite well and takes interfacial energies into account is given by (cf., [4,5])

$$F(\epsilon, \epsilon_x, \theta) = -c_e\,\theta\,\log(\theta/\theta_2) + c_e\,\theta + \bar{C} + \gamma(\theta - \theta_1)\epsilon^2 - \beta\epsilon^4 + \alpha\epsilon^6 + \delta\epsilon_x^2 \quad, \qquad (1.1)$$

where $c_e, \bar{C}, \theta_1, \theta_2, \alpha, \beta, \gamma, \delta$ denote physical constants. Note that in the range of interesting temperatures, for $\theta \to \theta_1$, F is not convex as function of ϵ.

[1]*Fachbereich 10 – Bauwesen, Universität–GH Essen, Postfach 10 37 64, D–4300 Essen 1, West–Germany; supported by DFG, SPP "Anwendungsbezogene Optimierung und Steuerung".*

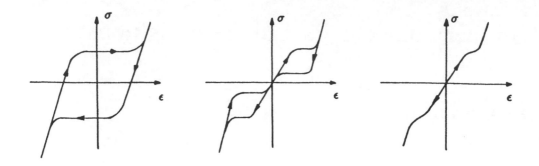

Figure 1: Typical stress–strain–curves in shape memory alloys, with temperature increasing from left to right.

The dynamics of thermomechanical processes in a solid are governed by the conservation laws of linear momentum, energy and mass. The latter may be ignored for the materials under consideration; the two others read

$$\rho u_{tt} - \sigma_x + \mu_{xx} = f, \quad \rho e_t + q_x - \sigma \epsilon_t - \mu \epsilon_{xt} = g,$$

where e – specific internal energy, q – heat flux, f – density of loads, g – density of sources/sinks. Assuming the Fourier form $q = -\kappa \theta_x$ and using the constitutive relations $\sigma = \frac{\partial F}{\partial \epsilon}$, $\mu = \frac{\partial F}{\partial \epsilon_x}$, $\rho e = F - \theta \frac{\partial F}{\partial \theta}$, we obtain for a sample of unit length the system

$$\rho u_{tt} - \left(2\gamma(\theta - \theta_1) - 4\beta u_x^3 + 6\alpha u_x^5\right)_x + 2\delta u_{xxxx} = f \quad , \tag{1.2a}$$

$$c_\epsilon \theta_t - 2\gamma \theta u_x u_{xt} - \kappa \theta_{xx} = g \quad , \tag{1.2b}$$

to be satisfied in $(0,1) \times (0,T)$, where $T > 0$ is some final time. In addition, we prescribe the initial and boundary conditions

$$u(x,0) = u_0(x),\ u_t(x,0) = u_1(x),\ \theta(x,0) = \theta_0(x), \quad x \in [0,1] \quad , \tag{1.2c}$$

$$u(0,t) = u_{xx}(0,t) = 0 = u(1,t) = u_{xx}(1,t), \quad t \in [0,T] \quad , \tag{1.2d}$$

$$\theta_x(0,t) = 0, \quad -\kappa \theta_x(1,t) = \bar{\kappa}\left(\theta(1,t) - \theta_\Gamma(t)\right), \quad t \in [0,T] \quad , \tag{1.2e}$$

where $\bar{\kappa} > 0$ is a heat exchange coefficient, and θ_Γ stands for the outside temperature at $x = 1$.

In the sequel, we state some results concerning well–posedness, optimal control and numerical approximation of the system (1.2a–e).

238

2 Well–Posedness

Consider (1.2a–e) with F given by (1.1). We assume

(H1) $u_0 \in \tilde{H}^4(0,1) := \{u \in H^4(0,1)|u(0) = u''(0) = 0 = u(1) = u''(1)\}$;

$u_1 \in \overset{\circ}{H}_1(0,1) \cap H^2(0,1)$; $\theta_0 \in H^2(0,1)$; $\theta_0(x) > 0$, $\quad \forall x \in [0,1]$.

(H2) $\theta_0'(0) = 0$; $\quad \theta_\Gamma^0 := \theta_0'(1) + \frac{\kappa}{\bar{\kappa}} \theta_0(1) > 0$.

Define the spaces

$$X := W^{2,\infty}(0,T;L^2(0,1)) \cap W^{1,\infty}(0,T;\overset{\circ}{H}_1(0,1) \cap H^2(0,1))$$
$$\cap L^\infty(0,T;\tilde{H}^4(0,1)) \quad, \tag{2.1a}$$
$$Y := H^1(0,T;H^1(0,1)) \cap L^2(0,T;H^3(0,1)) \quad, \tag{2.1b}$$
$$Z := H^1(0,T;H^1(0,1)) \times H^1(0,T;H^1(0,1)) \times H^1(0,T) \quad, \tag{2.1c}$$

and the sets

$$M_f := H^1(0,T;H^1(0,1)) \quad, \tag{2.2a}$$
$$M_g := \{g \in H^1(0,T;H^1(0,1))| \, g(x,t) \geq 0, \text{ on } [0,1] \times [0,T]\} \quad, \tag{2.2b}$$
$$M_{\theta_\Gamma} := \{\theta_\Gamma \in H^1(0,T)| \, \theta_\Gamma(0) = \theta_\Gamma^0, \, \theta_\Gamma(t) > 0 \text{ on } [0,T]\} \quad, \tag{2.2c}$$
$$M := M_f \times M_g \times M_{\theta_\Gamma} \quad. \tag{2.2d}$$

We have the result (cf., [9, Theorem 2.1])

Theorem 2.1 *Suppose (H1),(H2) hold. Then to every $(f,g,\theta_\Gamma) \in M$ the system (1.2a–e) has a unique solution $(u,\theta) \in X \times Y$ such that $\theta(x,t) > 0$ on $[0,1] \times [0,T]$. If, in addition, θ_0 satisfies compatibility conditions of sufficiently high order and if $u_0 \in H^5(0,1)$, $u_1 \in H^3(0,1)$, $\theta_0 \in H^4(0,1)$, $f_{tt} \in L^2(0,T;L^2(0,1))$, $g \in L^2(0,T;H^2(0,1))$, $\theta_\Gamma \in H^2(0,T)$, then the solution is classical and all the partial derivatives entering (1.2a,b) are Hölder continuous on $[0,1] \times [0,T]$.*

The next result is concerned with the differentiability properties of the solution operator S which assigns to each $(f,g,\theta_\Gamma) \in M$ the unique solution (u,θ). To this end, let $K \subset M$ denote some nonempty, closed and convex set. For $(f,g,\theta_\Gamma) \in K$ we define :

$$K^+(f,g,\theta_\Gamma) := \{(h,k,\ell) \in Z \,|\, (f + \lambda h, g + \lambda k, \theta_\Gamma + \lambda \ell) \in K$$
$$\text{for all sufficiently small } \lambda > 0\} \,. \tag{2.3}$$

The following result has been proved in [1] :

Theorem 2.2 *Suppose (H1),(H2) hold, and suppose $(f,g,\theta_\Gamma) \in K$. Then for any $(h,k,\ell) \in K^+(f,g,\theta_\Gamma)$ the operator S, viewed as mapping between M and the space*

$$B := (W^{1,\infty}(0,T;L^2(0,1)) \cap L^\infty(0,T;\overset{\circ}{H}_1(0,1) \cap H^2(0,1)))$$
$$\times (L^2(0,T;H^1(0,1)) \cap L^\infty(0,T;L^2(0,1))) \quad, \tag{2.4}$$

has a directional derivative at (f,g,θ_Γ) in the direction (h,k,ℓ). The directional derivative can be characterized as the solution of an associated linear initial–boundary value problem.

239

The above differentiability result can be used to derive first order necessary conditions of optimality for optimal control problems associated with the system (1.2a–e). In this connection it is natural to regard (f, g, θ_Γ) as control variables and to consider cost functionals involving the order parameter $\epsilon = u_x$. For details we refer to [1].

3 Numerical Approximation of Problem (1.2a–e)

To construct a convergent numerical scheme for the approximate solution of (1.2a–e), we choose $K, N, M \in \mathbb{N}$ and define

$$F_0(\epsilon, \theta) := \gamma(\theta - \theta_1)\epsilon^2 - \beta\epsilon^4 + \alpha\epsilon^6 \quad . \tag{3.1}$$

Moreover, we put $h = \frac{T}{M}$, $t_m^{(M)} = mh$, $0 \le m \le M$, and $x_i^{(N)} = \frac{i}{N}$, $0 \le i \le N$.

Let $Z_K = \text{span}\{z_1, \ldots, z_K\}$, where z_j denotes the j–th eigenfunction of the eigenvalue problem $z'''' = \lambda z$, in $(0,1)$, $z(0) = z''(0) = 0 = z(1) = z''(1)$, and denote by Y_N the linear space of linear splines on $[0,1]$ corresponding to the partition $\{x_i^{(N)}\}_{i=0}^N$ of $[0,1]$.

Now let P_K denote the H^4–orthogonal projection onto Z_K, Q_K the H^2–orthogonal projection onto Z_K, and R_N the H^1–orthogonal projection onto Y_N. Furthermore, we introduce the averages

$$f_M^m(x) = \frac{1}{h}\int_{(m-1)h}^{mh} f(x,t)dt, \quad g_M^m(x) = \frac{1}{h}\int_{(m-1)h}^{mh} g(x,t)dt,$$

$$\theta_{\Gamma,M}^m = \frac{1}{h}\int_{(m-1)h}^{mh} \theta_\Gamma(t)dt \quad . \tag{3.2}$$

We consider the discrete problem

$(D_{M,N,K})$:

Find $u^m = \sum_{k=1}^K \alpha_k^m z_k$, $\theta^m = \sum_{k=0}^N \beta_k^m y_k^{(N)}$, $1 \le m \le M$, such that

$$\int_0^1 (\rho \frac{u^m - 2u^{m-1} + u^{m-2}}{h^2}\xi + 2\delta u_{xx}^m \xi_{xx} - f_M^m \xi$$

$$+\xi_x (F_0(u_x^m, \theta^{m-1}) - F_0(u_x^{m-1}, \theta^{m-1}))/(u_x^m - u_x^{m-1}))\,dx = 0, \quad \forall \xi \in Z_K, \tag{3.3a}$$

$$\int_0^1 (c_e \frac{\theta^m - \theta^{m-1}}{h}\eta - \gamma\theta^{m-1}\frac{(u_x^m)^2 - (u_x^{m-1})^2}{h}\eta$$

$$+\kappa\theta_x^m \eta_x - g_M^m \eta)\,dx + \bar{\kappa}(\theta^m(1) - \theta_{\Gamma,M}^m)\eta(1) = 0, \quad \forall \eta \in Y_N, \tag{3.3b}$$

$$u^0 = P_K(u_0), \quad \frac{u^0 - u^{-1}}{h} = Q_K(u_1), \quad \theta^0 = R_N(\theta_0). \tag{3.3c}$$

The following result has been shown in [8] :

240

Theorem 3.1 *Suppose (H1),(H2) hold, and suppose N is sufficiently large. Then there exist constants $\bar{C}_1 > 0$, $\bar{C}_2 > 0$ which do not depend on M, N, K, such that for $\frac{c_e}{6\kappa N^2} < h \leq \bar{C}_1$ the discrete problem $(D_{M,N,K})$ has a solution which satisfies*

$$\theta^m(x) \geq 0, \quad \forall x \in [0,1], \quad 0 \leq m \leq M \quad , \tag{3.4a}$$

$$\max_{0 \leq m \leq M} \left(\| \frac{u^m - u^{m-1}}{h} \|^2 + \| \frac{u_x^m - u_x^{m-1}}{h} \|^2 + \|u_{xxx}^m\|^2 \right) \leq \bar{C}_2 \quad , \tag{3.4b}$$

$$\max_{0 \leq m \leq M} (\|\theta_x^m\|^2 + |\theta^m(1)|^2) + \sum_{m=1}^{M} h \| \frac{\theta^m - \theta^{m-1}}{h} \|^2 \leq \bar{C}_2 \quad . \tag{3.4c}$$

It is easy to obtain convergent approximations from Theorem 3.1. To this end, let $\varphi :$ $\mathbb{N} \to \mathbb{N}$ be strictly increasing. We take $N = \varphi(K)$, $M = M(N)$ with $\frac{T}{M} > \frac{c_e}{6\kappa N^2}$, and we choose $K \in \mathbb{N}$ large enough. Denoting the corresponding solutions of $(D_{M,N,K})$ by $\{(u_K^m, \theta_K^m)\}_{m=1}^{M}$ and introducing the linear–in–time interpolations

$$u_K(\cdot, t) = (Mt - m + 1)u_K^m + (m - Mt)u_K^{m-1} \quad ,$$
$$\theta_K(\cdot, t) = (Mt - m + 1)\theta_K^m + (m - Mt)\theta_K^{m-1} \quad ,$$
$$(m-1)h \leq t \leq mh \quad , \quad m = 1, \ldots, M \quad , \tag{3.5}$$

we obtain that (cf., [8])

$$u_{K,x} \to u_x, \quad \theta_K \to \theta, \text{ uniformly on } [0,1] \times [0,T]. \tag{3.6}$$

The above algorithm has been tested numerically for the alloy $Au_{23} Cu_{30} Zn_{47}$. For this alloy one has (cf., [4]):

$\alpha = 7.5 \times 10^6 \, J\,cm^{-3}$, $\beta = 1.5 \times 10^5 \, J\,cm^{-3}$, $\gamma = 24 \, J\,cm^{-3}\,K^{-1}$, $\delta = 10^{-12} \, J\,cm^{-1}$, $c_e = 2.9 \, J\,cm^{-3}\,K^{-1}$, $\theta_1 = 208 \, K$, $\kappa = 1.9 \, W\,cm^{-1}\,K^{-1}$, $\rho = 11.1 \, g\,cm^{-3}$.

The numerical values of $\bar{\kappa}$, f, u_1 were taken as zero; moreover, we chose $\theta_0(x) \equiv 200 \, K$, $h = 10^{-6} \, sec$, $N = 600$ and $K = 12$. As initial distribution of the displacement we took the H^4–orthogonal projektion onto Z_K of the function

$$\tilde{u}_0(x) = 0.118 \, x, \, x \leq \frac{1}{2}, \quad \tilde{u}_0(x) = 0.118 \, (1 - x), \, x \geq \frac{1}{2} \quad [cm] . \tag{3.7}$$

This means that initially, at $\theta_0 = 200 \, K$, we have an equilibrium configuration consisting of two distinct regions containing different types of martensite. We applied to the system the distributed heat pulse $g(x,t) \equiv 4 \times 10^6$ which was switched off after $10^{-4} \, secs$. The evolution of the system was followed until the final time $T = 10^{-3} \, secs$. During that period, the temperature was raised to about $309 \, K$ while the crystal lattice was completely transformed into the austenitic configuration. In Fig. 2 the evolutions of temperature and strain are displayed for the first $2 \times 10^{-4} \, secs$.

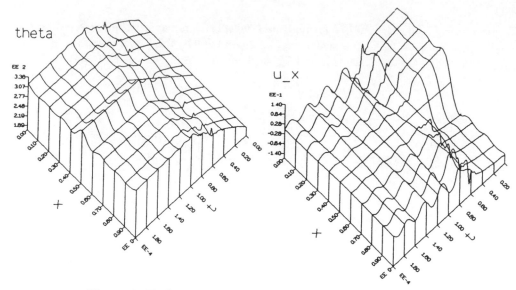

Figure 2: Evolution of temperature and strain distribution.

References

1. Brokate, M., Sprekels, J.: Optimal control of thermomechanical phase transitions in shape memory alloys: necessary conditions of optimality. Submitted.

2. Delaey, L., Chandrasekharan, M. (eds): International Conference on Martensitic Transformations, Proceedings, Les Editions de Physique, Les Ulis 1982.

3. Falk, F.: Landau theory and martensitic phase transitions, Journal de Physique C4, 12 (1982), 3–15.

4. Falk, F.: Ginzburg–Landau theory of static domain walls in shape–memory alloys. Phys. B–Condensed Matter 51 (1983), 177–185.

5. Falk, F.: Ginzburg–Landau theory and solitary waves in shape–memory alloys. Phys. B–Condensed Matter 54 (1984), 159–167.

6. Müller, I., Wilmański, K.: A model for phase transitions in pseudoelastic bodies. Il Nuovo Cimento 57B (1980), 283–318.

7. Müller, I., Wilmański, K.: Memory alloys – phenomenology and ersatzmodel. In: Brulin, O., Hsieh, R.K.T. (eds.), Continuum Models of Discrete Systems, Vol.IV, 495– 509, North–Holland, Amsterdam 1981.

8. Niezgódka, M., Sprekels, J.: Convergent numerical approximations of the thermo-mechanical phase transitions in shape memory alloys. Submitted.

9. Sprekels, J., Zheng, S.: Global solutions to the equations of a Ginzburg–Landau theory for structural phase transitions in shape memory alloys. Physica D 39 (1989), 59–76.